T0342130

PEARSON EDEXCEL INTERNATIONAL A LEVEL

PURE MATHEMATICS 4

Student Book

Series Editors: Joe Skrakowski and Harry Smith

Authors: Greg Attwood, Jack Barraclough, Ian Bettison, Lee Cope, Charles Garnet Cox, Keith Gallick, Daniel Goldberg, Alistair Macpherson, Anne McAteer, Lee McKelvey, Bronwen Moran, Su Nicholson, Diane Oliver, Laurence Pateman, Joe Petran, Keith Pledger, Cong San, Joe Skrakowski, Harry Smith, Geoff Staley, Robert Ward-Penny, Dave Wilkins

Published by Pearson Education Limited, 80 Strand, London, WC2R 0RL.

https://www.pearson.com/international-schools

Copies of official specifications for all Pearson qualifications may be found on the website: https://qualifications.pearson.com

Text © Pearson Education Limited 2019
Edited by Linnet Bruce
Typeset by Tech-Set Ltd, Gateshead, UK
Original illustrations © Pearson Education Limited 2019
Illustrated by © Tech-Set Ltd, Gateshead, UK
Cover design by © Pearson Education Limited 2019

Cover images: *Front*: **Getty Images:** Werner Van Steen
Inside front cover: **Shutterstock.com:** Dmitry Lobanov

The rights of Greg Attwood, Jack Barraclough, Ian Bettison, Lee Cope, Charles Garnet Cox, Keith Gallick, Daniel Goldberg, Alistair Macpherson, Anne McAteer, Lee McKelvey, Bronwen Moran, Su Nicholson, Diane Oliver, Laurence Pateman, Joe Petran, Keith Pledger, Cong San, Joe Skrakowski, Harry Smith, Geoff Staley, Robert Ward-Penny and Dave Wilkins to be identified as the authors of this work have been asserted by them in accordance with the Copyright, Designs and Patents Act 1988.

First published 2019

25 24
15 14 13

British Library Cataloguing in Publication Data
A catalogue record for this book is available from the British Library

ISBN 978 1 292245 12 6

Copyright notice
All rights reserved. No part of this may be reproduced in any form or by any means (including photocopying or storing it in any medium by electronic means and whether or not transiently or incidentally to some other use of this publication) without the written permission of the copyright owner, except in accordance with the provisions of the Copyright, Designs and Patents Act 1988 or under the terms of a licence issued by the Copyright Licensing Agency, 5th Floor, Shackleton House, 4 Battlebridge Lane, London, SE1 2HX (www.cla.co.uk). Applications for the copyright owner's written permission should be addressed to the publisher.

Printed by Neografia in Slovakia

Acknowledgements

Images:
Alamy Stock Photo: Terry Oakley 16; **Getty Images:** mikedabell 50, Westend61 97; **Science Photo Library:** Millard H. Sharp 66; **Shutterstock.com:** Karynav 6, LDprod 1, OliverSved 30

All other images © Pearson Education Limited 2019
All artwork © Pearson Education Limited 2019

Endorsement Statement
In order to ensure that this resource offers high-quality support for the associated Pearson qualification, it has been through a review process by the awarding body. This process confirms that this resource fully covers the teaching and learning content of the specification or part of a specification at which it is aimed. It also confirms that it demonstrates an appropriate balance between the development of subject skills, knowledge and understanding, in addition to preparation for assessment.

Endorsement does not cover any guidance on assessment activities or processes (e.g. practice questions or advice on how to answer assessment questions) included in the resource, nor does it prescribe any particular approach to the teaching or delivery of a related course.

While the publishers have made every attempt to ensure that advice on the qualification and its assessment is accurate, the official specification and associated assessment guidance materials are the only authoritative source of information and should always be referred to for definitive guidance.

Pearson examiners have not contributed to any sections in this resource relevant to examination papers for which they have responsibility.

Examiners will not use endorsed resources as a source of material for any assessment set by Pearson. Endorsement of a resource does not mean that the resource is required to achieve this Pearson qualification, nor does it mean that it is the only suitable material available to support the qualification, and any resource lists produced by the awarding body shall include this and other appropriate resources.

ABOUT THIS BOOK

The following three themes have been fully integrated throughout the Pearson Edexcel International Advanced Level in Mathematics series, so they can be applied alongside your learning.

1. Mathematical argument, language and proof

- Rigorous and consistent approach throughout
- Notation boxes explain key mathematical language and symbols

2. Mathematical problem-solving

- Hundreds of problem-solving questions, fully integrated into the main exercises
- Problem-solving boxes provide tips and strategies
- Challenge questions provide extra stretch

The Mathematical Problem-Solving Cycle

specify the problem → collect information → process and represent information → interpret results → (cycle)

3. Transferable skills

- Transferable skills are embedded throughout this book, in the exercises and in some examples
- These skills are signposted to show students which skills they are using and developing

Finding your way around the book

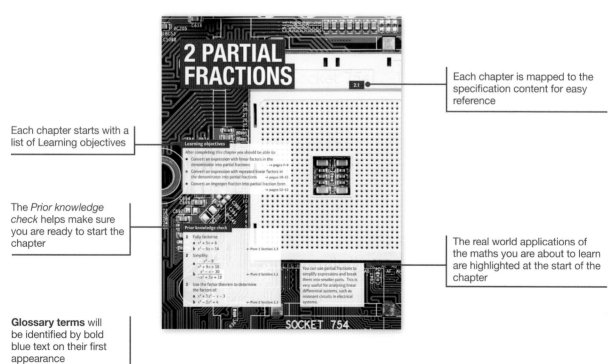

Each chapter is mapped to the specification content for easy reference

Each chapter starts with a list of Learning objectives

The *Prior knowledge check* helps make sure you are ready to start the chapter

The real world applications of the maths you are about to learn are highlighted at the start of the chapter

Glossary terms will be identified by bold blue text on their first appearance

Exercise questions are carefully graded to increase in difficulty and gradually bring you up to exam standard

Transferable skills are signposted where they naturally occur in the exercises and examples

Exercises are packed with exam-style questions to ensure you are ready for the exams

Exam-style questions are flagged with Ⓔ

Problem-solving questions are flagged with Ⓟ

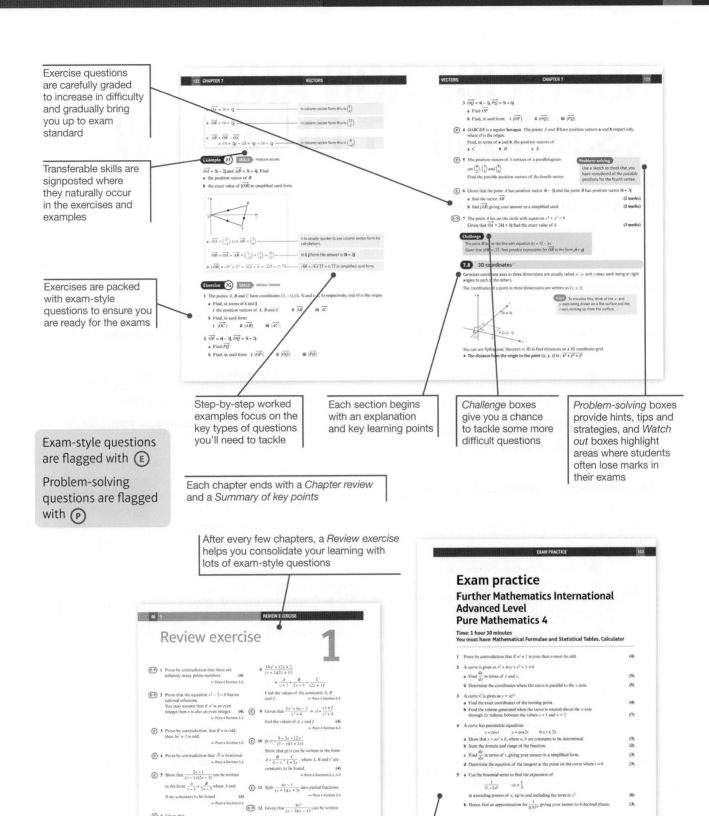

Step-by-step worked examples focus on the key types of questions you'll need to tackle

Each section begins with an explanation and key learning points

Challenge boxes give you a chance to tackle some more difficult questions

Problem-solving boxes provide hints, tips and strategies, and *Watch out* boxes highlight areas where students often lose marks in their exams

Each chapter ends with a *Chapter review* and a *Summary of key points*

After every few chapters, a *Review exercise* helps you consolidate your learning with lots of exam-style questions

A full practice paper at the back of the book helps you prepare for the real thing

QUALIFICATION AND ASSESSMENT OVERVIEW

Qualification and content overview

Pure Mathematics 4 (P4) is a **compulsory** unit in the following qualifications:

International Advanced Level in Mathematics

International Advanced Level in Pure Mathematics

Assessment overview

The following table gives an overview of the assessment for this unit.

We recommend that you study this information closely to help ensure that you are fully prepared for this course and know exactly what to expect in the assessment.

Unit	Percentage	Mark	Time	Availability
P4: Pure Mathematics 4 Paper code WMA14/01	$16\frac{2}{3}$ % of IAL	75	1 hour 30 mins	January, June and October First assessment June 2020

IAL: International Advanced A Level.

Assessment objectives and weightings

			Minimum weighting in IAS and IAL
AO1	Recall, select and use their knowledge of mathematical facts, concepts and techniques in a variety of contexts.		30%
AO2	Construct rigorous mathematical arguments and proofs through use of precise statements, logical deduction and inference and by the manipulation of mathematical expressions, including the construction of extended arguments for handling substantial problems presented in unstructured form.		30%
AO3	Recall, select and use their knowledge of standard mathematical models to represent situations in the real world; recognise and understand given representations involving standard models; present and interpret results from such models in terms of the original situation, including discussion of the assumptions made and refinement of such models.		10%
AO4	Comprehend translations of common realistic contexts into mathematics; use the results of calculations to make predictions, or comment on the context; and, where appropriate, read critically and comprehend longer mathematical arguments or examples of applications.		5%
AO5	Use contemporary calculator technology and other permitted resources (such as formulae booklets or statistical tables) accurately and efficiently; understand when not to use such technology, and its limitations. Give answers to appropriate accuracy.		5%

Relationship of assessment objectives to units

P4	Assessment objective				
	AO1	AO2	AO3	AO4	AO5
Marks out of 75	25–30	25–30	5–10	5–10	5–10
%	$33\frac{1}{3}$–40	$33\frac{1}{3}$–40	$6\frac{2}{3}$–$13\frac{1}{3}$	$6\frac{2}{3}$–$13\frac{1}{3}$	$6\frac{2}{3}$–$13\frac{1}{3}$

Calculators

Students may use a calculator in assessments for these qualifications. Centres are responsible for making sure that calculators used by their students meet the requirements given in the table below.

Students are expected to have available a calculator with at least the following keys: $+, -, \times, \div, \pi, x^2$, $\sqrt{x}, \frac{1}{x}, x^y$, ln x, e^x, $x!$, sine, cosine and tangent and their inverses in degrees and decimals of a degree, and in radians; memory.

Prohibitions

Calculators with any of the following facilities are prohibited in all examinations:

- databanks
- retrieval of text or formulae
- built-in symbolic algebra manipulations
- symbolic differentiation and/or integration
- language translators
- communication with other machines or the internet

Extra online content

Whenever you see an *Online* box, it means that there is extra online content available to support you.

SolutionBank

SolutionBank provides worked solutions for questions in the book. Download the solutions as a PDF or quickly find the solution you need online.

Use of technology

Explore topics in more detail, visualise problems and consolidate your understanding. Use pre-made GeoGebra activities or Casio resources for a graphic calculator.

Online Find the point of intersection graphically using technology.

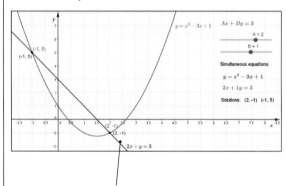

GeoGebra-powered interactives

Interact with the maths you are learning using GeoGebra's easy-to-use tools

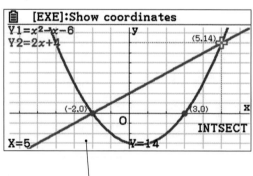

Graphic calculator interactives

Explore the maths you are learning and gain confidence in using a graphic calculator

Calculator tutorials

Our helpful video tutorials will guide you through how to use your calculator in the exams. They cover both Casio's scientific and colour graphic calculators.

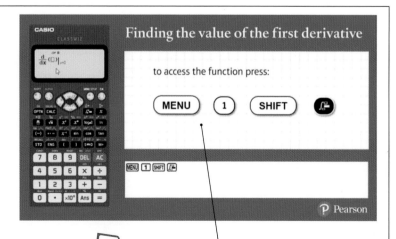

Online Work out each coefficient quickly using the ^nC_r and power functions on your calculator.

Step-by-step guide with audio instructions on exactly which buttons to press and what should appear on your calculator's screen

1 PROOF

Learning objectives

After completing this chapter you should be able to:

● Use proof by contradiction to prove true statements → **pages 2–5**

Prior knowledge check

1 Factorise:

  a $x^2 - 6x + 5$ **b** $x^2 - 16$

  c $9x^2 - 25$ ← **Pure 1 Section 1.3**

2 For any integers n and m, decide whether the following will always be odd, always be even, or could be either:

  a $8n$ **b** $n - m$

  c $3m$ **d** $2n - 5$

                     ← **International GCSE Mathematics**

You can use proof by contradiction to prove that there is an infinite number of prime numbers. Very large prime numbers are used to encode chip and pin transactions.

1.1 Proof by contradiction

A **contradiction** is a disagreement between two statements, which means that both cannot be true. Proof by contradiction is a powerful technique.

- To prove a statement by contradiction you start by assuming it is **not true**. You then use logical steps to show that this assumption leads to something impossible (either a contradiction of the assumption, or a contradiction of a fact you know to be true). You can conclude that your assumption was incorrect, and the original statement **was true**.

> **Notation** A statement that asserts the falsehood of another statement is called the negation of that statement.

Example **1** **SKILLS** REASONING

Prove by **contradiction** that there is no greatest odd **integer**.

Assumption: there is a greatest odd integer, n.	Begin by assuming the original statement is false. This is the **negation** of the original statement.
$n + 2$ is also an integer and $n + 2 > n$ $n + 2 = $ odd + even = odd	You need to use logical steps to reach a contradiction. Show all of your working.
So there exists an odd integer greater than n. This contradicts the assumption that the greatest odd integer is n.	The existence of an **odd** integer greater than n contradicts your initial assumption.
Therefore, there is no greatest odd integer.	Finish your proof by concluding that the original statement must be true.

Example **2** **SKILLS** REASONING

Prove by contradiction that if n^2 is even, then n must be even.

Assumption: There exists a number n such that n^2 is even and n is odd.	This is the negation of the original statement.
In this case integers j and k exist such that $n^2 = 2j$ and $n = 2k + 1$	You can write any even number in the form $2j$ and any odd number in the form $2k + 1$
Substituting for n we have: $(2k + 1)^2 = 2j$ $\Rightarrow 4k^2 + 4k + 1 = 2j$ $\Rightarrow j = \dfrac{4k^2 + 4k + 1}{2} = 2k^2 + 2k + \dfrac{1}{2}$ However, if k is an integer then j cannot be an integer, so there is a contradiction.	Carry out the substation and find j in terms of k.
Therefore our assumption that there exists a number n such that n^2 is even and n is odd is FALSE.	Finish your proof by concluding that the original statement must be false.

- A **rational** number can be written as $\dfrac{a}{b}$, where a and b are integers.

- An **irrational number** cannot be expressed in the form $\dfrac{a}{b}$, where a and b are integers.

> **Notation** $\mathbb{Q}$ is the set of all rational numbers.

Example 3

Prove by contradiction that $\sqrt{2}$ is an irrational number.

Assumption: $\sqrt{2}$ is a rational number.	Begin by assuming the original statement is false.
Then $\sqrt{2} = \dfrac{a}{b}$ for some integers, a and b.	This is the definition of a rational number.
Also assume that this fraction cannot be reduced further: there are no common factors between a and b.	If a and b did have a **common factor** you could just cancel until this **fraction** was in its simplest form.
So $2 = \dfrac{a^2}{b^2}$ or $a^2 = 2b^2$	**Square** both sides and make a^2 the subject.
This means that a^2 must be even, so a is also even.	We proved this result in Example 2.
If a is even, then it can be expressed in the form $a = 2n$, where n is an integer.	
So $a^2 = 2b^2$ becomes $(2n)^2 = 2b^2$ which means $4n^2 = 2b^2$ or $2n^2 = b^2$	
This means that b^2 must be even, so b is also even.	Again using the result from Example 2.
If a and b are both even, they will have a common factor of 2.	All even numbers are divisible by 2.
This contradicts the statement that a and b have no common factors.	
Therefore $\sqrt{2}$ is an irrational number.	Finish your proof by concluding that the original statement must be true.

Example 4   **SKILLS**   ADAPTIVE LEARNING

Prove by contradiction that there are **infinitely** many **prime numbers**.

Assumption: there is a finite number of prime numbers.	Begin by assuming the original statement is false.
List all the prime numbers that exist: $p_1, p_2, p_3, \ldots, p_n$	This is a list of all possible prime numbers.
Consider the number $N = p_1 \times p_2 \times p_3 \times \ldots \times p_n + 1$	This new number is one more than the **product** of the existing prime numbers.
When you divide N by any of the prime numbers $p_1, p_2, p_3, \ldots, p_n$ you get a remainder of 1.	
So none of the prime numbers $p_1, p_2, p_3, \ldots, p_n$ is a factor of N.	
So N must either be prime or have a prime factor which is not in the list of all possible prime numbers.	This contradicts the assumption that the list $p_1, p_2, p_3, \ldots, p_n$ contains all the prime numbers.
This is a contradiction.	
Therefore, there is an infinite number of prime numbers.	Conclude your proof by stating that the original statement must be true.

Exercise (1A) **SKILLS** REASONING

(P) **1** Select the statement that is the negation of 'All multiples of three are even'.
  A All multiples of three are odd.
  B At least one multiple of three is odd.
  C No multiples of three are even.

(P) **2** Write down the negation of each statement.
  a All rich people are happy.
  b There are no prime numbers between 10 million and 11 million.
  c If p and q are prime numbers then $(pq + 1)$ is a prime number.
  d All numbers of the form $2^n - 1$ are either prime numbers or multiples of 3.
  e At least one of the above four statements is true.

(P) **3** Statement: If n^2 is odd then n is odd.
  a Write down the negation of this statement.
  b Prove the original statement by contradiction.

(P) **4** Prove the following statements by contradiction.
  a There is no greatest even integer.
  b If n^3 is even then n is even.
  c If pq is even then at least one of p and q is even.
  d If $p + q$ is odd then at least one of p and q is odd.

(E/P) **5** **a** Prove that if ab is an irrational number then at least one of a and b is an irrational number.
  (3 marks)
  b Prove that if $a + b$ is an irrational number then at least one of a and b is an irrational number.
  (3 marks)
  c A student makes the following statement:
      If $a + b$ is a rational number then at least one of a and b is a rational number.
      Show by means of a counterexample that this statement is not true. **(1 mark)**

(P) **6** Use proof by contradiction to show that there exist no integers a and b for which $21a + 14b = 1$

> **Hint** Assume the opposite is true, and then divide both sides by the highest common **factor** of 21 and 14.

(E/P) **7** **a** Prove by contradiction that if n^2 is a multiple of 3, n is a multiple of 3. **(3 marks)**
  b Hence prove by contradiction that $\sqrt{3}$ is an irrational number. **(3 marks)**

> **Hint** Consider numbers in the form $3k + 1$ and $3k + 2$

(P) **8** Use proof by contradiction to prove the statement:
  'There are no integer solutions to the **equation** $x^2 - y^2 = 2$'

> **Hint** You can assume that x and y are **positive**, since $(-x)^2 = x^2$

(E/P) **9** Prove by contradiction that $\sqrt[3]{2}$ is irrational. **(5 marks)**

(P) **10** This student has attempted to use proof by contradiction to show that there is no least positive rational number:

> **Assumption:** There is a least positive rational number.
> Let this least positive rational number be n.
>
> As n is rational, $n = \dfrac{a}{b}$ where a and b are integers.
>
> $n - 1 = \dfrac{a}{b} - 1 = \dfrac{a - b}{b}$
>
> Since a and b are integers, $\dfrac{a - b}{b}$ is a rational number that is less than n.
>
> This contradicts the statement that n is the least positive rational number.
> Therefore, there is no least positive rational number.

Problem-solving

You might have to analyse student working like this in your exam. The question says, 'the error', so there should only be one error in the proof.

a Identify the **error** in the student's proof. **(1 mark)**

b Prove by contradiction that there is no least positive rational number. **(5 marks)**

Chapter review 1

1 Write down the negation of each statement.
   a At least two of the above statements are false.
   b People in cold countries are unhappy.
   c One quarter of the people who entered the competition won a prize.

2 Prove that if ab is rational, then no single number a or b can be irrational.

3 Select the statement that is the negation of 'All multiples of five are odd'.
   A At least one multiple of five is even.
   B No multiples of five are even.
   C All multiples of five are odd.

4 Prove by contradiction, that if $a - 2b$ is irrational, then at least one of a and b is an irrational number.

5 Use proof by contradiction to show that there are no integers x and y that can satisfy the equation $3x + 18y = 1$

6 Prove by contradiction that if n^4 is odd then n must be odd.

Summary of key points

1 To prove a statement by contradiction you start by assuming it is **not true**. You then use logical steps to show that this assumption leads to something impossible (either a contradiction of the assumption or a contradiction of a fact you know to be true). You can conclude that your assumption was incorrect, and the original statement **was true**.

2 A rational number can be written as $\dfrac{a}{b}$, where a and b are integers.

An irrational number cannot be expressed in the form $\dfrac{a}{b}$, where a and b are integers.

2 PARTIAL FRACTIONS

Learning objectives

After completing this chapter you should be able to:

● Convert an expression with linear factors in the
denominator into partial fractions → pages 7–9

● Convert an expression with repeated linear factors in
the denominator into partial fractions → pages 10–11

● Convert an improper fraction into partial fraction form
→ pages 12–13

Prior knowledge check

1 Fully factorise:
   a $x^2 + 5x + 6$
   b $x^2 - 5x - 14$ ← Pure 1 Section 1.3

2 Simplify:
   a $\dfrac{x^2 - 9}{x^2 + 9x + 18}$
   b $\dfrac{x^2 - x - 30}{-x^2 + 3x + 18}$ ← Pure 2 Section 1.1

3 Use the factor theorem to determine
   the factors of:
   a $x^3 + 3x^2 - x - 3$
   b $x^3 - 3x^2 + 4$ ← Pure 2 Section 1.3

You can use partial fractions to
simplify expressions and break
them into smaller parts. This is
very useful for analysing linear
differential systems, such as
resonant circuits in electrical
systems.

2.1 Partial fractions

- A single fraction with two **distinct linear** factors in the **denominator** can be split into two separate fractions with linear denominators. This is called splitting it into partial fractions.

$$\frac{5}{(x + 1)(x - 4)} \equiv \frac{A}{x + 1} + \frac{B}{x - 4}$$

A and *B* are **constants** to be found.

The **expression** is rewritten as the sum of two partial fractions.

The denominator contains two linear factors: $(x + 1)$ and $(x - 4)$

Links   Partial fractions are used for **binomial expansions** and **integration**.

There are two methods to find the constants *A* and *B*: by **substitution** and by **equating coefficients**.

Example **1**   **SKILLS**   **INTERPRETATION**

Split $\dfrac{6x - 2}{(x - 3)(x + 1)}$ into partial fractions by   **a** substitution   **b** equating coefficients.

a $\dfrac{6x - 2}{(x - 3)(x + 1)} \equiv \dfrac{A}{x - 3} + \dfrac{B}{x + 1}$

Set $\dfrac{6x - 2}{(x - 3)(x + 1)}$ **identical** to $\dfrac{A}{x - 3} + \dfrac{B}{x + 1}$

$\equiv \dfrac{A(x + 1) + B(x - 3)}{(x - 3)(x + 1)}$

Add the two fractions.

$6x - 2 \equiv A(x + 1) + B(x - 3)$

$6 \times 3 - 2 = A(3 + 1) + B(3 - 3)$

To find *A* substitute $x = 3$
This value of *x* **eliminates** *B* from the equation.

$16 = 4A$

$A = 4$

$6 \times (-1) - 2 = A(-1 + 1) + B(-1 - 3)$

$-8 = -4B$

To find *B* substitute $x = -1$
This value of *x* eliminates *A* from the equation.

$B = 2$

$\therefore \dfrac{6x - 2}{(x - 3)(x + 1)} \equiv \dfrac{4}{x - 3} + \dfrac{2}{x + 1}$

b $\dfrac{6x - 2}{(x - 3)(x + 1)} \equiv \dfrac{A}{x - 3} + \dfrac{B}{x + 1}$

$\equiv \dfrac{A(x + 1) + B(x - 3)}{(x - 3)(x + 1)}$

$6x - 2 \equiv A(x + 1) + B(x - 3)$

$\equiv Ax + A + Bx - 3B$

Expand the brackets.

$\equiv (A + B)x + (A - 3B)$

Collect like **terms**.

Equate coefficients of *x*:

$6 = A + B$ **(1)**

Equate constant terms:

$-2 = A - 3B$ **(2)**

You want $(A + B)x + A - 3B \equiv 6x - 2$
Hence coefficient of *x* is 6, and constant term is −2

(1) − (2):

$8 = 4B$

$\Rightarrow \quad B = 2$

Substitute $B = 2$ in **(1)** $\Rightarrow 6 = A + 2$

$A = 4$

Solve simultaneously.

■ The method of partial fractions can also be used when there are more than two distinct linear factors in the denominator.

For example, the expression $\dfrac{7}{(x-2)(x+6)(x+3)}$

can be split into $\dfrac{A}{x-2} + \dfrac{B}{x+6} + \dfrac{C}{x+3}$

The constants A, B and C can again be found either by substitution or by equating coefficients.

Watch out This method cannot be used for a repeated linear factor in the denominator. For example, the expression $\dfrac{9}{(x+4)(x-1)^2}$

cannot be rewritten as $\dfrac{A}{x+4} + \dfrac{B}{x-1} + \dfrac{C}{x-1}$

since $\dfrac{B}{x-1} + \dfrac{C}{x-1}$ is equivalent to $\dfrac{D}{x-1}$

which would leave you with the expression $\dfrac{A}{x+4} + \dfrac{D}{x-1}$

This is all due to the **repeated factor** $(x-1)$ There is more on this in the next section.

Example (**2**)   **SKILLS** ▷ PROBLEM-SOLVING

Given that $\dfrac{6x^2 + 5x - 2}{x(x-1)(2x+1)} \equiv \dfrac{A}{x} + \dfrac{B}{x-1} + \dfrac{C}{2x+1}$, find the values of the constants A, B and C.

Let $\dfrac{6x^2 + 5x - 2}{x(x-1)(2x+1)} \equiv \dfrac{A}{x} + \dfrac{B}{x-1} + \dfrac{C}{2x+1}$

The denominators must be x, $(x-1)$ and $(2x+1)$

$\equiv \dfrac{A(x-1)(2x+1) + Bx(2x+1) + Cx(x-1)}{x(x-1)(2x+1)}$

Add the fractions.

$\therefore\ 6x^2 + 5x - 2 \equiv A(x-1)(2x+1)$
$\qquad\qquad\qquad + Bx(2x+1) + Cx(x-1)$

The **numerators** are equal.

Let $x = 1$:

$\qquad 6 + 5 - 2 = 0 + B \times 1 \times 3 + 0$
$\qquad\qquad 9 = 3B$
$\qquad\qquad B = 3$

Let $x = 0$:

$\qquad 0 + 0 - 2 = A \times (-1) \times 1 + 0 + 0$
$\qquad\qquad -2 = -A$
$\qquad\qquad A = 2$

Proceed by substitution OR by equating coefficients.
Here we used the method of substitution.

Let $x = -\dfrac{1}{2}$:

$\qquad \dfrac{6}{4} - \dfrac{5}{2} - 2 = 0 + 0 + C \times \left(-\dfrac{1}{2}\right) \times \left(-\dfrac{3}{2}\right)$

$\qquad\qquad -3 = \dfrac{3}{4}C$

$\qquad\qquad C = -4$

So $\dfrac{6x^2 + 5x - 2}{x(x-1)(2x+1)} \equiv \dfrac{2}{x} + \dfrac{3}{x-1} - \dfrac{4}{2x+1}$

So $A = 2$, $B = 3$ and $C = -4$

Finish the question by listing the coefficients.

Exercise 2A **SKILLS** PROBLEM-SOLVING

1 Express as partial fractions

   a $\dfrac{6x - 2}{(x - 2)(x + 3)}$

   b $\dfrac{2x + 11}{(x + 1)(x + 4)}$

   c $\dfrac{-7x - 12}{2x(x - 4)}$

   d $\dfrac{2x - 13}{(2x + 1)(x - 3)}$

   e $\dfrac{6x + 6}{x^2 - 9}$

   Hint First **factorise** the denominator.

   f $\dfrac{7 - 3x}{x^2 - 3x - 4}$

   g $\dfrac{8 - x}{x^2 + 4x}$

   h $\dfrac{2x - 14}{x^2 + 2x - 15}$

(E) 2 Show that $\dfrac{-2x - 5}{(4 + x)(2 - x)}$ can be written in the form $\dfrac{A}{4 + x} + \dfrac{B}{2 - x}$ where A and B are constants to be found.   **(3 marks)**

(P) 3 The expression $\dfrac{A}{(x - 4)(x + 8)}$ can be written in partial fractions as $\dfrac{2}{x - 4} + \dfrac{B}{x + 8}$

Find the values of the constants A and B.

(E) 4 $h(x) = \dfrac{2x^2 - 12x - 26}{(x + 1)(x - 2)(x + 5)}, x > 2$

Given that $h(x)$ can be expressed in the form $\dfrac{A}{x + 1} + \dfrac{B}{x - 2} + \dfrac{C}{x + 5}$, find the values of A, B and C.   **(4 marks)**

(E) 5 Given that, for $x < -1$, $\dfrac{-10x^2 - 8x + 2}{x(2x + 1)(3x - 2)} \equiv \dfrac{D}{x} + \dfrac{E}{2x + 1} + \dfrac{F}{3x - 2}$, where D, E and F are constants. Find the values of D, E and F.   **(4 marks)**

6 Express as partial fractions

   a $\dfrac{2x^2 - 12x - 26}{(x + 1)(x - 2)(x + 5)}$

   b $\dfrac{-10x^2 - 8x + 2}{x(2x + 1)(3x - 2)}$

   c $\dfrac{-5x^2 - 19x - 32}{(x + 1)(x + 2)(x - 5)}$

(P) 7 Express as partial fractions

   a $\dfrac{6x^2 + 7x - 3}{x^3 - x}$

   b $\dfrac{8x + 9}{10x^2 + 3x - 4}$

   Hint First factorise the denominator.

Challenge

Express $\dfrac{5x^2 - 15x - 8}{x^3 - 4x^2 + x + 6}$ as a sum of fractions with linear denominators.

2.2 Repeated factors

- A single fraction with a repeated linear factor in the denominator can be split into two or more separate fractions.

In this case, there is a special method for dealing with the repeated linear factor.

A and B and C are constants to be found.

$$\frac{2x + 9}{(x - 5)(x + 3)^2} \equiv \frac{A}{x - 5} + \frac{B}{x + 3} + \frac{C}{(x + 3)^2}$$

The denominator contains three linear factors: $(x - 5)$, $(x + 3)$ and $(x + 3)$. $(x + 3)$ is a repeated linear factor.

The expression is rewritten as the sum of three partial fractions. Notice that $(x - 5)$, $(x + 3)$ and $(x + 3)^2$ are the denominators.

Example 3 **SKILLS** **PROBLEM-SOLVING**

Show that $\dfrac{11x^2 + 14x + 5}{(x + 1)^2(2x + 1)}$ can be written in the form $\dfrac{A}{x + 1} + \dfrac{B}{(x + 1)^2} + \dfrac{C}{2x + 1}$, where A, B and C are constants to be found.

Let
$$\frac{11x^2 + 14x + 5}{(x + 1)^2(2x + 1)} \equiv \frac{A}{(x + 1)} + \frac{B}{(x + 1)^2} + \frac{C}{(2x + 1)}$$

You need denominators of $(x + 1)$, $(x + 1)^2$ and $(2x + 1)$

$$\equiv \frac{A(x + 1)(2x + 1) + B(2x + 1) + C(x + 1)^2}{(x + 1)^2(2x + 1)}$$

Add the three fractions.

Hence $11x^2 + 14x + 5$
$$\equiv A(x + 1)(2x + 1) + B(2x + 1) + C(x + 1)^2 \quad (1)$$

The numerators are equal.

Let $x = -1$:
$$11 - 14 + 5 = A \times 0 + B \times -1 + C \times 0$$
$$2 = -1B$$
$$B = -2$$

To find B substitute $x = -1$

Let $x = -\dfrac{1}{2}$:
$$\frac{11}{4} - 7 + 5 = A \times 0 + B \times 0 + C \times \frac{1}{4}$$

To find C substitute $x = -\dfrac{1}{2}$

$$\frac{3}{4} = \frac{1}{4}C$$
$$C = 3$$
$$11 = 2A + C$$

Equate terms in x^2 in (1). Terms in x^2 are $A \times 2x^2 + C \times x^2$

$$11 = 2A + 3$$
$$2A = 8$$
$$A = 4$$

Substitute $C = 3$

Hence $\dfrac{11x^2 + 14x + 5}{(x + 1)^2(2x + 1)}$

Finish the question by listing the coefficients.

$$\equiv \frac{4}{(x + 1)} - \frac{2}{(x + 1)^2} + \frac{3}{(2x + 1)}$$

So $A = 4$, $B = -2$ and $C = 3$

Online Check your answer using the simultaneous equations function on your calculator.

Exercise **2B**　　**SKILLS**　　PROBLEM-SOLVING

(E) **1** $f(x) = \dfrac{3x^2 + x + 1}{x^2(x + 1)}, x \neq 0, x \neq -1$

Given that $f(x)$ can be expressed in the form $\dfrac{A}{x} + \dfrac{B}{x^2} + \dfrac{C}{x + 1}$, find the values of

A, B and C.　　　　　　　　　　　　　　　　　　　　　　　**(4 marks)**

(E) **2** $g(x) = \dfrac{-x^2 - 10x - 5}{(x + 1)^2(x - 1)}, x \neq -1, x \neq 1$

Find the values of the constants D, E and F such that $g(x) = \dfrac{D}{x + 1} + \dfrac{E}{(x + 1)^2} + \dfrac{F}{x - 1}$　　**(4 marks)**

(E) **3** Given that, for $x < 0$, $\dfrac{2x^2 + 2x - 18}{x(x - 3)^2} \equiv \dfrac{P}{x} + \dfrac{Q}{x - 3} + \dfrac{R}{(x - 3)^2}$, where P, Q and R are constants,

find the values of P, Q and R.　　　　　　　　　　　　　　　　　**(4 marks)**

(E) **4** Show that $\dfrac{5x^2 - 2x - 1}{x^3 - x^2}$ can be written in the form $\dfrac{C}{x} + \dfrac{D}{x^2} + \dfrac{E}{x - 1}$ where C, D and E

are constants to be found.　　　　　　　　　　　　　　　　　　　**(4 marks)**

(E) **5** $p(x) = \dfrac{2x}{(x + 2)^2}, x \neq -2$

Find the values of the constants A and B such that $p(x) = \dfrac{A}{x + 2} + \dfrac{B}{(x + 2)^2}$　　　**(4 marks)**

(E) **6** $\dfrac{10x^2 - 10x + 17}{(2x + 1)(x - 3)^2} \equiv \dfrac{A}{2x + 1} + \dfrac{B}{x - 3} + \dfrac{C}{(x - 3)^2}, x > 3$

Find the values of the constants A, B and C.　　　　　　　　　　　　　**(4 marks)**

(E) **7** Show that $\dfrac{39x^2 + 2x + 59}{(x + 5)(3x - 1)^2}$ can be written in the form $\dfrac{A}{x + 5} + \dfrac{B}{3x - 1} + \dfrac{C}{(3x - 1)^2}$

where A, B and C are constants to be found.　　　　　　　　　　　　　**(4 marks)**

(P) **8** Express as partial fractions:

a $\dfrac{4x + 1}{x^2 + 10x + 25}$　　　　　　b $\dfrac{6x^2 - x + 2}{4x^3 - 4x^2 + x}$

2.3 Improper fractions

- An improper fraction is one that is top heavy, where the power of the denominator is equal to or greater than the power of the numerator. An improper fraction can be split into partial fractions.

Some fractions can be given as $\dfrac{x^2}{(x+1)(x-1)}$. Since the top and bottom are both quadratics in this case, dividing one by the other should produce a constant, so the form would be $A + \dfrac{B}{x+1} + \dfrac{C}{x-1}$.

It is similar if the expression $\dfrac{x^3}{(x+1)(x-1)}$ is split into partial fractions. Then, by first noting the difference in powers between numerator and denominator, this can be written in the form $Ax + B + \dfrac{C}{x+1} + \dfrac{D}{x-1}$. This is because a cubic over a quadratic produces a linear function.

Example 4

Express $\dfrac{2x}{x+1}$ as partial fractions.

$$\frac{2x}{x+1} = A + \frac{B}{x+1}$$

Set $\dfrac{2x}{x+1}$ identical to $A + \dfrac{B}{(x+1)}$ since linear over linear must be a constant.

$$= \frac{A(x+1)+B}{x+1}$$

Add the fractions.

$$2x = A(x+1)+B$$

$$2 \times (-1) = A(-1+1)+B$$

To find B, substitute in $x = -1$
This value eliminates A from the equation.

$$B = -2$$

$$2 \times 0 = A(0+1)-2$$

To find A use any value except -2.

$$A = 2$$

$$\frac{2x}{x+1} = 2 - \frac{2}{x+1}$$

Example 5

Express $\dfrac{3x^3}{(x-1)(x-2)}$ as partial fractions.

$$\frac{3x^3}{(x-1)(x-2)} = Ax + B + \frac{C}{x-1} + \frac{D}{x-2}$$

Set $\dfrac{3x^3}{(x-1)(x-2)}$ identical to

$Ax + B + \dfrac{C}{x-1} + \dfrac{D}{x-2}$ since cubic over quadratic must be linear.

Add the fractions.

$$= \frac{(Ax + B)(x - 1)(x - 2) + C(x - 2) + D(x - 1)}{(x - 1)(x - 2)}$$

To determine A, note that $\dfrac{3x^3}{x^2 + \ldots}$ must be $3x + B$

$$= \frac{(3x + B)(x - 1)(x - 2) + C(x - 2) + D(x - 1)}{(x - 1)(x - 2)}$$

$3x^3 = (3x + B)(x - 1)(x - 2) + C(x - 2) + D(x - 1)$

To find C, substitute in $x = 1$, this value eliminates B and D from the equation.

$3 \times 1^3 = (3 \times 1 + B)(1 - 1)(1 - 2) + C(1 - 2) + D(1 - 1)$

$3 = -C$

$C = -3$

$3 \times 2^3 = (3 \times 2 + B)(2 - 1)(2 - 2) - 3(2 - 2) + D(2 - 1)$

To find D, substitute in $x = 2$, this value eliminates B and C from the equation.

$24 = D$

$3 \times 0^3 = (3 \times 0 + B)(0 - 1)(0 - 2) - 3(0 - 2) + 24(0 - 1)$

To find B, substitute in $x = 0$

$0 = 2B + 6 - 24$

$B = 9$

Exercise 2C

1 Express as partial fractions

  a $\dfrac{x + 1}{x + 2}$ **b** $\dfrac{4 - x}{x + 1}$ **c** $\dfrac{x^2 + 1}{x - 1}$ **d** $\dfrac{2x^2}{x + 3}$

2 Show that $\dfrac{x^2}{(x - 2)(x + 3)}$ can be written in the form $A + \dfrac{B}{x - 2} + \dfrac{C}{x + 3}$, where A, B and C are constants to be found.

3 Given that $\dfrac{Ax^2 + Bx}{(x - 1)(x + 1)}$ can be expressed in the form $2 + \dfrac{5}{2(x - 1)} + \dfrac{C}{2(x + 1)}$, find the values of A, B and C.

4 Express as partial fractions

  a $\dfrac{x^2 - 1}{x + 3}$ **b** $\dfrac{2x^2 - 2}{x(x + 3)}$ **c** $\dfrac{2 - 3x^2}{(2x - 1)(x + 1)}$

5 Given that $\dfrac{x^3}{(x + 2)(x - 1)}$ can be expressed in the form $Ax + B + \dfrac{C}{x + 2} + \dfrac{D}{x - 1}$, find the values of A, B, C and D.

6 Express as partial fractions

  a $\dfrac{1 + x^3}{x(x + 2)}$ **b** $\dfrac{x^3 - x}{(x + 2)(x - 2)}$

7 Given that $\dfrac{x^2}{(x + 1)^2}$ can be expressed in the form $A + \dfrac{B}{x + 1} + \dfrac{C}{(x + 1)^2}$, find the values of A, B and C.

8 Express as partial fractions

  a $\dfrac{x^2 + 1}{(x - 2)^2}$ **b** $\dfrac{2x^3 - 1}{(x + 2)^2}$

Challenge

Express $\dfrac{3x^3}{(x-1)^3}$ as partial fractions.

Chapter review (2)

1. Given that $\dfrac{4}{(x+1)(x-4)}$ can be written in the form $\dfrac{A}{x+1} + \dfrac{B}{x-4}$, find the values of A and B.

2. Express as partial fractions

   a $\dfrac{8x+13}{(x+2)(x+1)}$ **b** $\dfrac{3-x}{(x-1)(x+3)}$

3. Show that $\dfrac{x}{(x+1)(x-2)(x+5)}$ can be written in form $\dfrac{A}{x+1} + \dfrac{B}{x-2} + \dfrac{C}{x+5}$, giving the values of A, B and C.

4. Express as partial fractions

   a $\dfrac{3x^2+7x-2}{x(x+1)(x-1)}$ **b** $\dfrac{6x^2-7x-18}{(x^2-4)(x-3)}$ **c** $\dfrac{x^2}{(x+1)(x+2)(x+3)}$

5. By first using the factor theorem to simplify the denominator, write $\dfrac{3x-2}{x^3+x^2-14x-24}$ as partial fractions.

6. Express as partial fractions

   a $\dfrac{2}{x^3-2x^2-x+2}$ **b** $\dfrac{3x+1}{x^3+5x^2+6x}$

7. Express as partial fractions

   a $\dfrac{x}{(x-3)^2(x+1)}$ **b** $\dfrac{3}{x^2(x+1)}$

8. Given that $\dfrac{2x}{(x+3)^2}$ can be written in the form $\dfrac{A}{x+3} + \dfrac{B}{(x+3)^2}$ determine the values of A and B.

9. Express $\dfrac{4}{x^3-3x+2}$ as partial fractions.

10. Express as partial fractions

    a $\dfrac{3x-1}{x+4}$ **b** $\dfrac{x^2+1}{x+2}$

11. Show that $\dfrac{x^2+2}{(x-2)^2}$ can be written in the form $A + \dfrac{B}{x-2} + \dfrac{C}{(x-2)^2}$ giving the values of A, B and C.

12. Express as partial fractions

    a $\dfrac{3-x^2}{(x+1)(x-2)}$ **b** $\dfrac{4x^2}{x-4}$

13. Given that $\dfrac{x^3}{(x+3)^2}$ can be written as $Ax + B + \dfrac{C}{x+3} + \dfrac{D}{(x+3)^2}$ find the values of A, B, C and D.

Summary of key points

1 A single fraction with two distinct linear factors in the denominator can be split into two separate fractions with linear denominators. This is called splitting it into partial fractions:

$$\frac{5}{(x+1)(x-4)} = \frac{A}{(x+1)} + \frac{B}{(x-4)}$$

2 The method of partial fractions can also be used when there are more than two distinct linear factors in the denominator:

$$\frac{7}{(x-2)(x+6)(x+3)} = \frac{A}{(x-2)} + \frac{B}{(x+6)} + \frac{C}{(x+3)}$$

3 A single fraction with a repeated linear factor in the denominator can be split into two or more separate fractions:

$$\frac{2x+9}{(x-5)(x+3)^2} = \frac{A}{(x-5)} + \frac{B}{(x+3)} + \frac{C}{(x+3)^2}$$

4 An improper algebraic fraction is one whose numerator has a degree equal to or larger than the denominator. An improper fraction must be converted to a mixed fraction before you can express it in partial fractions.

To convert an improper fraction into a mixed fraction you can use:

- algebraic division
- or the relationship $F(x) = Q(x) \times$ divisor + remainder

3 COORDINATE GEOMETRY IN THE (x, y) PLANE

Learning objectives

After completing this chapter you should be able to:

* Convert parametric equations into Cartesian form by substitution → **pages 17–18**
* Convert parametric equations into Cartesian form using trigonometric identities → **pages 21–23**
* Understand and use parametric equations of curves and sketch parametric curves → **pages 25–26**
* Solve coordinate geometry problems involving parametric equations → **pages 17–26**

Prior knowledge check

1 Rearrange to make t the subject:

   a $x = 4t - kt$ **b** $y = 3t^2$ **c** $y = 2 - 4\ln t$ **d** $x = 1 + 2e^{-3t}$

   ← **Pure 3 Sections 4.2, 4.4**

2 Write in terms of powers of $\cos x$:

   a $4 + 3\sin^2 x$ **b** $\sin 2x$ **c** $\cot x$ **d** $2\cos x + \cos 2x$

   ← **Pure 2 Section 6.3**
   ← **Pure 3 Sections 3.1, 3.3, 3.4**

3 State the ranges of the following functions:

   a $y = \ln(x + 1), x > 0$ **b** $y = 2\sin x, 0 < x < \pi$

   c $y = x^2 + 4x - 2, -4 < x < 1$ **d** $y = \dfrac{1}{2x + 5}, x > -2$

   ← **Pure 1 Sections 2.4, 4.2**
   ← **Pure 2 Section 6.1**
   ← **Pure 3 Section 4.4**

4 A circle has centre $(0, 4)$ and radius 5. Find the coordinates of the points of intersection of the circle and the line with equation $2y - x - 10 = 0$

   ← **Pure 2 Section 2.3**

Parametric equations are an alternative coordinate system to Cartesian equations. They can be used to represent an otherwise complicated Cartesian equation in a simpler, and more accessible, form.

3.1 Parametric equations

You can write the x- and y-**coordinates** of each point on a **curve** as functions of a third **variable**. This variable is called a parameter and is often represented by the letter t.

■ A curve can be defined using parametric equations $x = p(t)$ and $y = q(t)$. Each value of the parameter, t, defines a point on the curve with coordinates $(p(t), q(t))$.

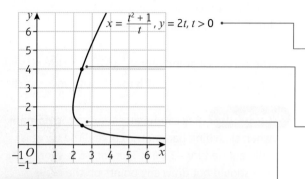

$x = \dfrac{t^2 + 1}{t},\ y = 2t,\ t > 0$

These are the parametric equations of the curve. The **domain** of the parameter tells you the values of t you would need to substitute to find the coordinates of the points on the curve.

When $t = 2$, $x = \dfrac{2^2 + 1}{2} = 2.5$ and $y = 2 \times 2 = 4$

This corresponds to the point $(2.5, 4)$

When $t = 0.5$, $x = \dfrac{0.5^2 + 1}{0.5} = 2.5$ and $y = 2 \times 0.5 = 1$

This corresponds to the point $(2.5, 1)$

Watch out The value of the parameter t is generally not equal to either the x- or the y-coordinate, and more than one point on the curve can have the same x-coordinate.

■ You can convert between parametric equations and **Cartesian** equations by using substitution to eliminate the parameter.

Notation A Cartesian equation in two **dimensions** involves the variables x and y only.

You can use the domain and range of the parametric functions to find the domain and range of the resulting Cartesian function.

■ For parametric equations $x = p(t)$ and $y = q(t)$ with Cartesian equation $y = f(x)$:
  • the domain of $f(x)$ is the range of $p(t)$
  • the range of $f(x)$ is the range of $q(t)$

Example ①   **SKILLS** ▸ ANALYSIS

A curve has parametric equations

$$x = 2t, \quad y = t^2, \quad -3 < t < 3$$

a Find a Cartesian equation of the curve in the form $y = f(x)$

b State the domain and range of $f(x)$

c **Sketch** the curve within the given domain for t.

---

a $x = 2t$ so $t = \dfrac{x}{2}$ (1)

            $y = t^2$ (2)

   Substitute (1) into (2):

   $y = \left(\dfrac{x}{2}\right)^2 = \dfrac{x^2}{4}$

A Cartesian equation only involves the variables x and y, so you need to eliminate t.

Rearrange one equation into the form $t = \ldots$ then substitute into the other equation.

This is a **quadratic** curve.

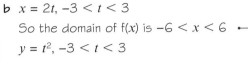

b $x = 2t, \; -3 < t < 3$

So the domain of f(x) is $-6 < x < 6$

$y = t^2, \; -3 < t < 3$

So the range of f(x) is $0 \leqslant y < 9$

The domain of f is the range of the parametric function for x. The range of $x = 2t$ over the domain $-3 < t < 3$ is $-6 < x < 6$

The range of f is the range of the parametric function for y. Choose your inequalities carefully. $y = t^2$ can equal 0 in the interval $-3 < t < 3$ so use $\leqslant$ but it cannot equal 9, so use $<$

c

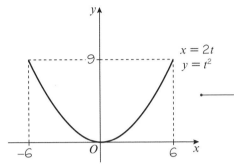

The curve is a graph of $y = \frac{1}{4}x^2$
Use your answers to part **b** to help with your sketch.

Watch out   Pay careful attention to the **domain** when sketching parametric curves. The curve is only defined for $-3 < t < 3$, or for $-6 < x < 6$ You should not draw any points on the curve outside that range.

Example **2**   **SKILLS**   CRITICAL THINKING

A curve has parametric equations

$$x = \ln(t + 3), \quad y = \frac{1}{t + 5}, \quad t > -2$$

a Find a Cartesian equation of the curve of the form $y = f(x)$, $x > k$ where k is a constant to be found.

b Write down the range of f(x).

a $x = \ln(t + 3)$

$e^x = t + 3$

So $e^x - 3 = t$

Substitute $t = e^x - 3$ into

$$y = \frac{1}{t + 5} = \frac{1}{e^x - 3 + 5} = \frac{1}{e^x + 2}$$

When $t = -2$: $x = \ln(t + 3) = \ln 1 = 0$

As t increases $\ln(t + 3)$ increases, so the range of the parametric function for x is $x > 0$

The Cartesian equation is

$$y = \frac{1}{e^x + 2}, \; x > 0$$

b When $t = -2$: $y = \frac{1}{t + 5} = \frac{1}{3}$

As t increases y decreases, but is always positive, so the range of the parametric function for y is $0 < y < \frac{1}{3}$

The range of f(x) is $0 < y < \frac{1}{3}$

Online   Sketch this parametric curve using technology.

e^x is the **inverse** function of $\ln x$.

Rearrange the equation for x into the form $t = \ldots$ then substitute into the equation for y.

To find the domain for f(x), consider the range of values x can take for values of $t > -2$

You need to consider what value x takes when $t = -2$ **and** what happens when t **increases**.

The range of f is the range of values y can take within the given range of the parameter.

You could also find the range of f(x) by considering the domain of f(x). $f(0) = \frac{1}{3}$ and f(x) **decreases** as x increases, so $y < \frac{1}{3}$

← Pure 1 Section 2.3

Exercise **3A**   **SKILLS** CRITICAL THINKING

1 Find a **Cartesian** equation for each of these **parametric equations**, giving your answer in the form $y = f(x)$. In each case find the domain and range of $f(x)$.

  a $x = t - 2, \quad y = t^2 + 1, \quad -4 \leqslant t \leqslant 4$ **b** $x = 5 - t, \quad y = t^2 - 1, \quad t \in \mathbb{R}$

  c $x = \dfrac{1}{t}, \quad y = 3 - t, \quad t \neq 0$

> **Notation** If the domain of t is given as $t \neq 0$, this implies that t can take any value in $\mathbb{R}$ other than 0.

  d $x = 2t + 1, \quad y = \dfrac{1}{t}, \quad t > 0$ **e** $x = \dfrac{1}{t - 2}, \quad y = t^2, \quad t > 2$

  f $x = \dfrac{1}{t + 1}, \quad y = \dfrac{1}{t - 2}, \quad t > 2$

2 For each of these parametric curves
  i find a Cartesian equation for the curve in the form $y = f(x)$ giving the domain on which the curve is defined
  ii find the range of $f(x)$.

  a $x = 2\ln(5 - t), \quad y = t^2 - 5, \quad t < 4$ **b** $x = \ln(t + 3), \quad y = \dfrac{1}{t + 5}, \quad t > -2$

  c $x = e^t, \quad y = e^{3t}, \quad t \in \mathbb{R}$

P **3** A curve C is defined by the parametric equations $x = \sqrt{t}, \quad y = t(9 - t), \quad 0 \leqslant t \leqslant 5$

  a Find a Cartesian equation of the curve in the form $y = f(x)$ and **determine** the domain and range of $f(x)$.

  b Sketch C showing clearly the coordinates of any turning points, **endpoints** and **intersections** with the coordinate **axes**.

> **Problem-solving**
>
> $y = t(9 - t)$ is a quadratic with a **negative** t^2 term and **roots** at $t = 0$ and $t = 9$. It will take its **maximum** value when $t = 4.5$

4 For each of the following parametric curves:
  i find a Cartesian equation for the curve in the form $y = f(x)$
  ii find the domain and range of $f(x)$
  iii sketch the curve within the given domain of t.

  a $x = 2t^2 - 3, \quad y = 9 - t^2, \quad t > 0$ **b** $x = 3t - 1, \quad y = (t - 1)(t + 2), \quad -4 < t < 4$

  c $x = t + 1, \quad y = \dfrac{1}{t - 1}, \quad t \in \mathbb{R}, \quad t \neq 1$ **d** $x = \sqrt{t} - 1, \quad y = 3\sqrt{t}, \quad t > 0$

  e $x = \ln(4 - t), \quad y = t - 2, \quad t < 3$

(P) **5** The curves C_1 and C_2 are defined by the following parametric equations.

$$C_1: \quad x = 1 + 2t, \quad y = 2 + 3t \quad 2 < t < 5 \qquad\qquad C_2: \quad x = \frac{1}{2t-3}, \quad y = \frac{t}{2t-3} \quad 2 < t < 3$$

a Show that both curves are **segments** of the same straight line.

b Find the length of each line segment.

> **Notation** Straight lines and line segments can be referred to as 'curves' in coordinate geometry.

(E/P) **6** A curve C has parametric equations

$$x = \frac{3}{t} + 2, \quad y = 2t - 3 - t^2, \quad t \in \mathbb{R}, \quad t \neq 0$$

a Determine the ranges of x and y in the given domain of t. **(3 marks)**

b Show that the Cartesian equation of C can be written in the form

$$y = \frac{A(x^2 + bx + c)}{(x - 2)^2}$$

where A, b and c are integers to be determined. **(3 marks)**

7 A curve has parametric equations

$$x = \ln(t + 3), \quad y = \frac{1}{t + 5}, \quad t > -2$$

a Show that a Cartesian equation of this curve is $y = f(x)$, $x > k$ where k is a constant to be found.

b Write down the range of $f(x)$.

(E/P) **8** A diagram shows a curve C with parametric equations

$$x = 3\sqrt{t}, \quad y = t^3 - 2t, \quad 0 \leqslant t \leqslant 2$$

a Find a Cartesian equation of the curve in the form $y = f(x)$, and state the domain of $f(x)$. **(3 marks)**

b Show that $\dfrac{dy}{dx} = 0$ when $t = \sqrt{\dfrac{2}{3}}$ **(3 marks)**

c Hence determine the range of $f(x)$. **(2 marks)**

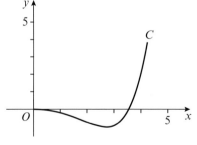

(E/P) **9** A curve C has parametric equations

$$x = t^3 - t, \quad y = 4 - t^2, \quad t \in \mathbb{R}$$

a Show that the Cartesian equation of C can be written in the form

$$x^2 = (a - y)(b - y)^2$$

where a and b are integers to be determined. **(3 marks)**

b Write down the maximum value of the y-coordinate for any point on this curve. **(2 marks)**

Challenge

A curve C has parametric equations

$$x = \frac{1 - t^2}{1 + t^2}, \quad y = \frac{2t}{1 + t^2}, \quad t \in \mathbb{R}$$

a Show that a Cartesian equation for this curve is $x^2 + y^2 = 1$

b Hence describe C.

3.2 Using trigonometric identities

You can use trigonometric identities to convert trigonometric parametric equations into Cartesian form. In this chapter you will always consider angles measured in **radians**.

Example ③　**SKILLS**　ADAPTIVE INNOVATION

A curve has parametric equations $x = \sin t + 2, \quad y = \cos t - 3, \quad t \in \mathbb{R}$

a Show that a Cartesian equation of the curve is $(x - 2)^2 + (y + 3)^2 = 1$

b Hence sketch the curve.

a $x = \sin t + 2$

So $\sin t = x - 2$　　**(1)**

$y = \cos t - 3$

$\cos t = y + 3$　　**(2)**

Substitute **(1)** and **(2)** into

$\sin^2 t + \cos^2 t \equiv 1$

$(x - 2)^2 + (y + 3)^2 = 1$

b

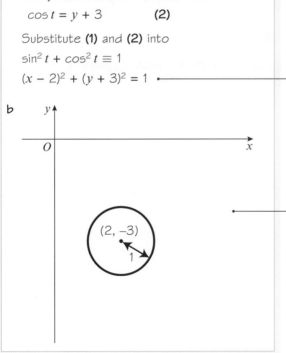

Problem-solving

If you can write expressions for $\sin t$ and $\cos t$ in terms of x and y then you can use the **identity** $\sin^2 t + \cos^2 t \equiv 1$ to eliminate the parameter, t.

← **Pure 2 Section 6.3**

Your equations in (1) and (2) are in terms of $\sin t$ and $\cos t$ so you need to square them when you substitute. Make sure you square the whole expression.

$(x - a)^2 + (y - b)^2 = r^2$ is the equation of a **circle** with centre (a, b) and **radius** r.

So the curve is a circle with centre $(2, -3)$ and radius 1.　　← **Pure 2 Section 2.2**

Example **4** SKILLS CREATIVITY

A curve is defined by the parametric equations

$$x = \sin t, \quad y = \sin 2t, \quad -\frac{\pi}{2} \leqslant t \leqslant \frac{\pi}{2}$$

a Find a Cartesian equation of the curve in the form

$$y = f(x), \quad -k \leqslant x \leqslant k$$

stating the value of the constant k.

b Write down the range of $f(x)$.

Online You can graph the parametric equations using technology.

a $y = \sin 2t$

$\quad = 2 \sin t \cos t$

$\quad = 2x \cos t$ (1)

$\sin^2 t + \cos^2 t \equiv 1$

$\cos^2 t \equiv 1 - \sin^2 t$

$\quad = 1 - x^2$

$\cos t = \sqrt{1 - x^2}$ (2)

Substitute (2) into (1): $y = 2x\sqrt{1 - x^2}$

When $t = -\dfrac{\pi}{2}$, $x = \sin\left(-\dfrac{\pi}{2}\right) = -1$

When $t = \dfrac{\pi}{2}$, $x = \sin\left(\dfrac{\pi}{2}\right) = 1$

The Cartesian equation is $y = 2x\sqrt{1 - x^2}$,

$-1 \leqslant x \leqslant 1$ so $k = 1$

b $-1 \leqslant y \leqslant 1$

Use the identity $\sin 2t \equiv 2 \sin t \cos t$, then substitute $x = \sin t$

Use the identity $\sin^2 t + \cos^2 t \equiv 1$ together with $x = \sin t$ to find an expression for $\cos t$ in terms of x.

Watch out Be careful when taking square roots. In this case you don't need to consider the negative **square root** because $\cos t$ is positive for all values in the domain of the parameter.

To find the domain of $f(x)$, consider the range of $x = \sin t$ for the values of the parameter given.

Within $-\dfrac{\pi}{2} \leqslant t \leqslant \dfrac{\pi}{2}$, $y = \sin 2t$ takes a minimum value of -1 and a maximum value of 1.

Example **5** SKILLS CREATIVITY

A curve C has parametric equations

$$x = \cot t + 2 \quad y = \operatorname{cosec}^2 t - 2, \quad 0 < t < \pi$$

a Find the equation of the curve in the form $y = f(x)$ and state the domain of x for which the curve is defined.

b Hence, sketch the curve.

a $x = \cot t + 2$

$\cot t = x - 2$ **(1)**

$y = \operatorname{cosec}^2 t - 2$

$\operatorname{cosec}^2 t = y + 2$ **(2)**

Substitute **(1)** and **(2)** into
$1 + \cot^2 t \equiv \operatorname{cosec}^2 t$

    $1 + (x - 2)^2 = y + 2$

    $1 + x^2 - 4x + 4 = y + 2$

           $y = x^2 - 4x + 3$

The range of $x = \cot t + 2$ over the domain $0 < t < \pi$ is all of the real numbers, so the domain of $f(x)$ is $x \in \mathbb{R}$

b $y = x^2 - 4x + 3 = (x - 3)(x - 1)$ is a quadratic with roots at $x = 3$ and $x = 1$ and y-intercept 3. The minimum point is $(2, -1)$

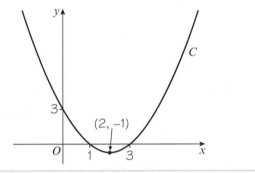

$(2, -1)$

Problem-solving

The parametric equations involve $\cot t$ and $\operatorname{cosec}^2 t$ so you can use the identity
$1 + \cot^2 t \equiv \operatorname{cosec}^2 t$ ← **Pure 3 Section 3.4**

Rearrange to find expressions for $\cot t$ and $\operatorname{cosec}^2 t$ in terms of x and y.

Expand and rearrange to make y the subject. You could also write the equation as:
$y = (x - 2)^2 - 1$
This is the completed square form which is useful when sketching the curve.

Consider the range of values taken by x over the domain of the parameter. The curve is defined on all of the real numbers, so it is the whole quadratic curve

Online Explore this curve graphically using technology.

Exercise   **SKILLS**   PROBLEM-SOLVING

1 Find the Cartesian equation of the curves given by the following parametric equations.

  a $x = 2\sin t - 1$, $\quad y = 5\cos t + 4$, $\quad 0 < t < 2\pi$
             b $x = \cos t$, $\quad y = \sin 2t$, $\quad 0 < t < 2\pi$

  c $x = \cos t$, $\quad y = 2\cos 2t$, $\quad 0 < t < 2\pi$
             d $x = \sin t$, $\quad y = \tan 2t$, $\quad 0 < t < \dfrac{\pi}{2}$

  e $x = \cos t + 2$, $\quad y = 4\sec t$, $\quad 0 < t < \dfrac{\pi}{2}$
     f $x = 3\cot t$, $\quad y = \operatorname{cosec} t$, $\quad 0 < t < \pi$

2 A circle has parametric equations $x = \sin t - 5$, $\quad y = \cos t + 2$

  a Find a Cartesian equation of the circle.

  b Write down the radius and the coordinates of the centre of the circle.

  c Write down a suitable domain of t which defines one full **revolution** around the circle.

Problem-solving

Think about how x and y change as t varies.

3 A circle has parametric equations $x = 4\sin t + 3$, $\quad y = 4\cos t - 1$

  Find the radius and the coordinates of the centre of the circle.

4 A curve is given by the parametric equation $x = \cos t - 2$, $y = \sin t + 3$, $-\pi < t < \pi$
Sketch the curve.

P **5** Find the Cartesian equation of the curves given by the following parametric equations.

a $x = \sin t$, $y = \sin\left(t + \dfrac{\pi}{4}\right)$, $-\dfrac{\pi}{2} < t < \dfrac{\pi}{2}$

b $x = 3\cos t$, $y = 2\cos\left(t + \dfrac{\pi}{6}\right)$, $0 < t < \dfrac{\pi}{3}$

> **Hint**   Use the addition **formulae** and **exact** values.

c $x = \sin t$, $y = 3\sin(t + \pi)$, $0 < t < 2\pi$

E **6** The curve C has parametric equations
$$x = 8\cos t, \quad y = \frac{1}{4}\sec^2 t, \quad -\frac{\pi}{2} < t < \frac{\pi}{2}$$

a Find a Cartesian equation of C.   **(4 marks)**

b Sketch the curve C on the appropriate domain.   **(3 marks)**

E **7** A curve has parametric equations
$$x = 3\cot^2 2t, \quad y = 3\sin^2 2t, \quad 0 < t \leqslant \frac{\pi}{4}$$

Find a Cartesian equation of the curve in the form $y = f(x)$

State the domain on which $f(x)$ is defined.   **(6 marks)**

E/P **8** A curve C has parametric equations
$$x = \frac{1}{3}\sin t, \quad y = \sin 3t, \quad 0 < t < \frac{\pi}{2}$$

a Show that the Cartesian equation of the curve is given by
$$y = ax(1 - bx^2)$$
where a and b are integers to be found.   **(5 marks)**

b State the domain and range of $y = f(x)$ in the given domain of t.   **(2 marks)**

E/P **9** Show that the curve with parametric equations
$$x = 2\cos t, \quad y = \sin\left(t - \frac{\pi}{6}\right), \quad 0 < t < \pi$$

can be written in the form
$$y = \frac{1}{4}\left(\sqrt{12 - 3x^2} - x\right), -2 < x < 2$$ **(6 marks)**

E/P **10** A curve has parametric equations
$$x = \tan^2 t + 5, \quad y = 5\sin t, \quad 0 < t < \frac{\pi}{2}$$

a Find the Cartesian equation of the curve in the form $y^2 = f(x)$.   **(4 marks)**

b Determine the possible values of x and y in the given domain of t.   **(2 marks)**

(E/P) **11** A curve C has parametric equations

$$x = \tan t, \quad y = 3 \sin(t - \pi), \quad 0 < t < \frac{\pi}{2}$$

Find a Cartesian equation of C. **(4 marks)**

> **Challenge**
>
> The curve C is given by the parametric equations:
>
> $$x = \frac{1}{2}\cos 2t, \quad y = \sin\left(t + \frac{\pi}{6}\right), \quad 0 < t < 2\pi$$
>
> Show that a Cartesian equation for C is $(4y^2 - 2 + 2x)^2 + 12x^2 - 3 = 0$

3.3 Curve sketching

Most parametric curves do not result in curves you will recognise and can sketch easily. You can **plot** any parametric curve by substituting values of the parameter into each equation.

Example 6   SKILLS   INTERPRETATION

Draw the curve given by parametric equations

$$x = 3\cos t + 4, \quad y = 2\sin t, \quad 0 \leqslant t \leqslant 2\pi$$

t	O	$\frac{\pi}{4}$	$\frac{\pi}{2}$	$\frac{3\pi}{4}$	π	$\frac{5\pi}{4}$	$\frac{3\pi}{2}$	$\frac{7\pi}{4}$	2π
$x = 3\cos t + 4$	7	6.12	4	1.88	1	1.88	4	6.12	7
$y = 2\sin t$	O	1.41	2	1.41	O	−1.41	−2	−1.41	O

This parametric curve has Cartesian equation

$$\left(\frac{x-4}{3}\right)^2 + \left(\frac{y}{2}\right)^2 = 1$$

This isn't a form of curve that you need to be able to recognise, but you can plot the curve using a table of values.

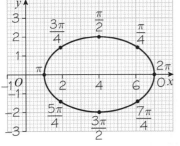

Choose values for t covering the domain of t. For each value of t, substitute to find **corresponding** values for x and y which will be the coordinates of points on the curve.

Plot the points and draw the curve through the points. The curve is an ellipse.

Example (7)   SKILLS   INTERPRETATION

Draw the curve given by the parametric equations $x = 2t, \quad y = t^2, \quad -1 \le t \le 5$

t	−1	0	1	2	3	4	5
$x = 2t$	−2	0	2	4	6	8	10
$y = t^2$	1	0	1	4	9	16	25

Online Use technology to graph the parametric equations.

Only calculate values of x and y for values of t in the given domain.

This is a 'partial' graph of the quadratic equation

$$y = \frac{x^2}{4}$$

You could also plot this curve by converting to Cartesian form and considering the domain of and range of the Cartesian function.

The domain is $-2 \le x \le 10$ and the range is $0 \le y \le 25$

Exercise (3C)   SKILLS   INTERPRETATION

1 A curve is given by the parametric equations

$$x = 2t, \quad y = \frac{5}{t}, \quad t \ne 0$$

Copy and complete the table and draw a graph of the curve for $-5 \le t \le 5$

t	−5	−4	−3	−2	−1	−0.5	0.5	1	2	3	4	5
$x = 2t$	−10	−8				−1						
$y = \dfrac{5}{t}$	−1	−1.25					10					

2 A curve is given by the parametric equations

$$x = t^2, \quad y = \frac{t^3}{5}$$

Copy and complete the table and draw a graph of the curve for $-4 \le t \le 4$

t	−4	−3	−2	−1	0	1	2	3	4
$x = t^2$	16								
$y = \dfrac{t^3}{5}$	−12.8								

3 A curve is given by parametric equations

$$x = \tan t + 1, \quad y = \sin t, \quad -\frac{\pi}{4} \le t \le \frac{\pi}{3}$$

Copy and complete the table and draw a graph of the curve for the given domain of t.

t	$-\dfrac{\pi}{4}$	$-\dfrac{\pi}{6}$	$-\dfrac{\pi}{12}$	0	$\dfrac{\pi}{12}$	$\dfrac{\pi}{6}$	$\dfrac{\pi}{4}$	$\dfrac{\pi}{3}$
$x = \tan t + 1$	0			1				
$y = \sin t$				0				

4 Draw the curves given by these parametric equations:

a $x = t - 2$, $y = t^2 + 1$, $-4 \leqslant t \leqslant 4$

b $x = 3\sqrt{t}$, $y = t^3 - 2t$, $0 \leqslant t \leqslant 2$

c $x = t^2$, $y = (2 - t)(t + 3)$, $-5 \leqslant t \leqslant 5$

d $x = 2 \sin t - 1$, $y = 5 \cos t + 1$, $-\dfrac{\pi}{4} \leqslant t \leqslant \dfrac{\pi}{4}$

e $x = \sec^2 t - 3$, $y = 2 \sin t + 1$, $-\dfrac{\pi}{4} \leqslant t \leqslant \dfrac{\pi}{2}$

f $x = t - 3 \cos t$, $y = 1 + 2 \sin t$, $0 \leqslant t \leqslant 2\pi$

(E) **5** The curve C has parametric equations

$$x = 3 - t, \quad y = t^2 - 2, \quad -2 \leqslant t \leqslant 3$$

a Find a Cartesian equation of C in the form $y = f(x)$ **(4 marks)**

b Draw the curve C on the appropriate domain.   **(3 marks)**

(E/P) **6** The curve C has parametric equations

$$x = 9 \cos t - 2, \quad y = 9 \sin t + 1, \quad -\dfrac{\pi}{6} \leqslant t \leqslant \dfrac{\pi}{2}$$

a Show that the Cartesian equation of C can be written as

$$(x + a)^2 + (y + b)^2 = c$$

where a, b and c are integers to be determined.   **(4 marks)**

b Draw the curve C on the given domain of t.   **(3 marks)**

c Find the length of C.   **(2 marks)**

> **Challenge**
>
> Sketch the curve given by the parametric equations on the given domain of t:
>
> $$x = \dfrac{9t}{1 + t^3}, \quad y = \dfrac{9t^2}{1 + t^3}, \quad t \neq -1$$
>
> Comment on the behaviour of the curve as t approaches -1 from the positive **direction** and from the negative direction.

Chapter review (3)

1 The diagram shows a sketch of the curve with parametric equations

$$x = 4\cos t, \quad y = 3\sin t, \quad 0 \leqslant t < 2\pi$$

a Find the coordinates of the points A and B.

b The point C has parameter $t = \dfrac{\pi}{6}$

Find the exact coordinates of C.

c Find the Cartesian equation of the curve.

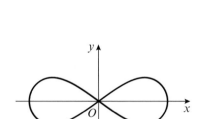

2 The diagram shows a sketch of the curve with parametric equations

$$x = \cos t, \quad y = <\frac{1}{2}\sin 2t, \quad 0 \leqslant t < 2\pi$$

The curve is symmetrical about both axes.

Copy the diagram and label the points having

parameters $t = 0$, $t = \dfrac{\pi}{2}$, $t = \pi$ and $t = \dfrac{3\pi}{2}$

(P) 3 A curve has parametric equations

$$x = e^{2t+1} + 1, \quad y = t + \ln 2, \quad t > 1$$

a Find a Cartesian equation of this curve in the form $y = f(x)$, $x > k$ where k is a constant to be found in exact form.

b Write down the range of $f(x)$, leaving your answer in exact form.

4 A curve has parametric equations

$$x = \frac{1}{2t + 1}, \quad y = 2\ln\left(t + \frac{1}{2}\right), \quad t > \frac{1}{2}$$

Find a Cartesian equation of the curve in the form $y = f(x)$, and state the domain and range of $f(x)$.

5 A circle has parametric equations $x = 4\sin t - 3, \quad y = 4\cos t + 5, \quad 0 \leqslant t \leqslant 2\pi$

a Find a Cartesian equation of the circle.

b Draw a sketch of the circle.

c Find the exact coordinates of the points of intersection of the circle with the y-axis.

(E/P) 6 The curve C has parametric equations

$$x = \frac{2 - 3t}{1 + t}, \quad y = \frac{3 + 2t}{1 + t}, \quad 0 \leqslant t \leqslant 4$$

a Show that the curve C is part of a straight line. **(3 marks)**

b Find the length of this line segment. **(2 marks)**

(E) **7** A curve C has parametric equations

$$x = t^2 - 2, \quad y = 2t, \quad 0 \leqslant t \leqslant 2$$

  a Find the Cartesian equation of C in the form $y = f(x)$ **(3 marks)**

  b State the domain and range of $y = f(x)$ in the given domain of t.   **(3 marks)**

  c Sketch the curve in the given domain of t.   **(2 marks)**

(E/P) **8** A curve C has parametric equations

$$x = 2\cos t, \quad y = 2\sin t - 5, \quad 0 \leqslant t \leqslant \pi$$

  a Show that the curve C forms part of a circle.   **(3 marks)**

  b Sketch the curve in the given domain of t.   **(3 marks)**

  c Find the length of the curve in the given domain of t.   **(3 marks)**

(E/P) **9** The curve C has parametric equations

$$x = t - 2, \quad y = t^3 - 2t^2, \quad t \in \mathbb{R}$$

  a Find a Cartesian equation of C in the form $y = f(x)$ **(3 marks)**

  b Sketch the curve C.   **(3 marks)**

Summary of key points

1 A curve can be defined using parametric equations $x = p(t)$ and $y = q(t)$
Each value of the parameter t, defines a point on the curve with coordinates $(p(t), q(t))$

2 You can convert between parametric equations and Cartesian equations by using substitution to eliminate the parameter.

3 For parametric equations $x = p(t)$ and $y = q(t)$ with Cartesian equation $y = f(x)$

  • the domain of $f(x)$ is the range of $p(t)$

  • the range of $f(x)$ is the range of $q(t)$

4 BINOMIAL EXPANSION

Learning objectives

After completing this chapter you should be able to:

● Expand $(1 + x)^n$ for any rational constant n and determine the range of values of x for which the expansion is valid
→ **pages 31–34**

● Expand $(a + bx)^n$ for any rational constant n and determine the range of values of x for which the expansion is valid
→ **pages 36–38**

● Use partial fractions to expand fractional expressions
→ **pages 40–41**

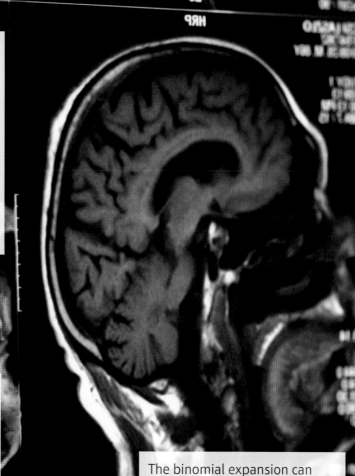

Prior knowledge check

1 Expand the following expressions in ascending powers of x up to and including the term in x^3:

   a $(1 + 5x)^7$ **b** $(5 - 2x)^{10}$ **c** $(1 - x)(2 + x)^6$

   ← **Pure 2 Section 4.3**

2 Write each of the following using partial fractions:

   a $\dfrac{-14x + 7}{(1 + 2x)(1 - 5x)}$ **b** $\dfrac{24x - 1}{(1 + 2x)^2}$

   c $\dfrac{24x^2 + 48x + 24}{(1 + x)(4 - 3x)^2}$

   ← **Pure 4 Sections 2.1, 2.2**

The binomial expansion can be used to find polynomial approximations for expressions involving fractional and negative indices. Medical physicists use these approximations to analyse magnetic fields in an MRI scanner.

4.1 Expanding $(1 + x)^n$

If n is a natural number you can find the binomial expansion for $(a + bx)^n$ using the formula:

$$(a + b)^n = a^n + \binom{n}{1}a^{n-1}b + \binom{n}{2}a^{n-2}b^2 + \ldots + \binom{n}{r}a^{n-r}b^r + \ldots + b^n, \; (n \in \mathbb{N})$$

Hint There are $n + 1$ terms, so this formula produces a **finite** number of terms.

If n is a **fraction** or a **negative number** you need to use a different version of the binomial expansion.

- This form of the binomial expansion can be applied to negative or fractional values of n to **obtain** an infinite **series**.

$$(1 + x)^n = 1 + nx + \frac{n(n-1)}{2!}x^2 + \frac{n(n-1)(n-2)}{3!}x^3 + \ldots + \left(\frac{n(n-1)\ldots(n-r+1)}{r!}\right)x^r + \ldots$$

- The expansion is valid when $|x| < 1, \, n \in \mathbb{R}$

When n is not a natural number, none of the factors in the expression $n(n-1) \ldots (n-r+1)$ are equal to zero. This means that this version of the binomial expansion produces an **infinite number** of terms.

Watch out This expansion is valid for any **real value** of n, but is **only** valid for values of x that satisfy $|x| < 1$, or in other words, when $-1 < x < 1$

Example 1   SKILLS   PROBLEM-SOLVING

Find the first four terms in the binomial expansion of $\dfrac{1}{1 + x}$

$$\frac{1}{1 + x} = (1 + x)^{-1}$$

Write in **index** form.

$$= 1 + (-1)x + \frac{(-1)(-2)x^2}{2!}$$

$$+ \frac{(-1)(-2)(-3)x^3}{3!} + \ldots$$

Replace n by -1 in the expansion.

$$= 1 - 1x + 1x^2 - 1x^3 + \ldots$$

As n is not a positive integer, no coefficient will ever be equal to zero. Therefore, the expansion is **infinite**.

$$= 1 - x + x^2 - x^3 + \ldots$$

For the series to be **convergent**, $|x| < 1$

- The expansion of $(1 + bx)^n$, where n is negative or a fraction, is valid for $|bx| < 1$, or $|x| < \dfrac{1}{|b|}$

Example **2**   **SKILLS**   **PROBLEM-SOLVING**

Find the binomial expansions of

a $(1 - x)^{\frac{1}{3}}$

b $\dfrac{1}{(1 + 4x)^2}$

up to and including the term in x^3. State the range of values of x for which each expansion is valid.

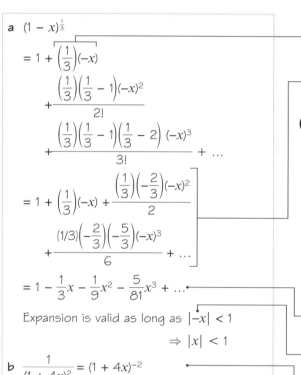

a $(1 - x)^{\frac{1}{3}}$

$= 1 + \left(\dfrac{1}{3}\right)(-x)$

$\quad + \dfrac{\left(\dfrac{1}{3}\right)\left(\dfrac{1}{3} - 1\right)(-x)^2}{2!}$

$\quad + \dfrac{\left(\dfrac{1}{3}\right)\left(\dfrac{1}{3} - 1\right)\left(\dfrac{1}{3} - 2\right)(-x)^3}{3!} + \dots$

$= 1 + \left(\dfrac{1}{3}\right)(-x) + \dfrac{\left(\dfrac{1}{3}\right)\left(-\dfrac{2}{3}\right)(-x)^2}{2}$

$\quad + \dfrac{(1/3)\left(-\dfrac{2}{3}\right)\left(-\dfrac{5}{3}\right)(-x)^3}{6} + \dots$

$= 1 - \dfrac{1}{3}x - \dfrac{1}{9}x^2 - \dfrac{5}{81}x^3 + \dots$

Expansion is valid as long as $|-x| < 1$

$\Rightarrow |x| < 1$

b $\dfrac{1}{(1 + 4x)^2} = (1 + 4x)^{-2}$

$= 1 + (-2)(4x)$

$\quad + \dfrac{(-2)(-2 - 1)(4x)^2}{2!}$

$\quad + \dfrac{(-2)(-2 - 1)(-2 - 2)(4x)^3}{3!} + \dots$

$= 1 + (-2)(4x)$

$\quad + \dfrac{(-2)(-3)16x^2}{2}$

$\quad + \dfrac{(-2)(-3)(-4)64x^3}{6} + \dots$

$= 1 - 8x + 48x^2 - 256x^3 + \dots$

Expansion is valid as long as $|4x| < 1$

$\Rightarrow |x| < \dfrac{1}{4}$

Replace n by $\dfrac{1}{3}$ and x by $(-x)$.

Simplify brackets.

Watch out   Be careful working out whether each term should be positive or negative:
- even number of negative signs means term is positive
- odd number of negative signs means term is negative

The x^3 term here has 5 negative signs in total, so it is negative.

Simplify coefficients.

Terms in expansion are $(-x)$, $(-x)^2$, $(-x)^3$

Write in index form.

Replace n by -2, x by $4x$

Simplify brackets.

Simplify coefficients.

Terms in expansion are $(4x)$, $(4x)^2$, $(4x)^3$

Online   Use technology to explore why the expansions are only valid for certain values of x.

Example (3) **SKILLS** ANALYSIS

a Find the expansion of $\sqrt{1 - 2x}$ up to and including the term in x^3.

b By substituting in $x = 0.01$, find a decimal approximation to $\sqrt{2}$.

a $\sqrt{1 - 2x} = (1 - 2x)^{\frac{1}{2}}$ —————— Write in index form.

$$= 1 + \left(\frac{1}{2}\right)(-2x)$$

$$+ \frac{\left(\frac{1}{2}\right)\left(\frac{1}{2} - 1\right)(-2x)^2}{2!}$$ ——— Replace n by $\frac{1}{2}$ and x by $(-2x)$

$$+ \frac{\left(\frac{1}{2}\right)\left(\frac{1}{2} - 1\right)\left(\frac{1}{2} - 2\right)(-2x)^3}{3!} + \dots$$

$$= 1 + \left(\frac{1}{2}\right)(-2x)$$

$$+ \frac{\left(\frac{1}{2}\right)\left(-\frac{1}{2}\right)(4x^2)}{2!}$$ ——— Simplify brackets.

$$+ \frac{\left(\frac{1}{2}\right)\left(-\frac{1}{2}\right)\left(-\frac{3}{2}\right)(-8x^3)}{6} + \dots$$

$$= 1 - x - \frac{x^2}{2} - \frac{x^3}{2} + \dots$$ ——— Simplify coefficients.

Expansion is valid if $|-2x| < 1$ ——— Terms in expansion are $(-2x)$, $(-2x)^2$, $(-2x)^3$

$$\Rightarrow |x| < \frac{1}{2}$$ ——— $x = 0.01$ satisfies the validity **condition** $|x| < \frac{1}{2}$

b $\sqrt{1 - 2 \times 0.01} \approx 1 - 0.01 - \frac{0.01^2}{2}$

$$- \frac{0.01^3}{2}$$ ——— Substitute $x = 0.01$ into both sides of the expansion.

$\sqrt{0.98} \approx 1 - 0.01 - 0.00005$

$$- 0.0000005$$ ——— Simplify both sides. Note that the terms are getting smaller.

$\sqrt{\dfrac{98}{100}} \approx 0.9899495$ ——— Write 0.98 as $\frac{98}{100}$

$\sqrt{\dfrac{49 \times 2}{100}} \approx 0.9899495$

$\dfrac{7\sqrt{2}}{10} \approx 0.9899495$ ——— Use rules of **surds**.

$\sqrt{2} \approx \dfrac{0.9899495 \times 10}{7}$ ——— This approximation is **accurate** to 7 decimal places.

$\sqrt{2} \approx 1.414213571$

Example **4** **SKILLS** CRITICAL THINKING

$f(x) = \dfrac{2 + x}{\sqrt{1 + 5x}}$

a Find the x^2 term in the series expansion of f(x).

b State the range of values of x for which the expansion is valid.

---

a f(x) = $(2 + x)(1 + 5x)^{-\frac{1}{2}}$ •————————— Write in index form.

$(1 + 5x)^{-\frac{1}{2}} = 1 + \left(-\dfrac{1}{2}\right)(5x)$

$+ \dfrac{\left(-\dfrac{1}{2}\right)\left(-\dfrac{3}{2}\right)}{2!}(5x)^2$

$+ \dfrac{\left(-\dfrac{1}{2}\right)\left(-\dfrac{3}{2}\right)\left(-\dfrac{5}{2}\right)}{3!}(5x)^3 + \ldots$ •————————— Find the binomial expansion of $(1 + 5x)^{-\frac{1}{2}}$

$= 1 - \dfrac{5}{2}x + \dfrac{75}{8}x^2 - \dfrac{625}{16}x^3 + \ldots$ •————————— Simplify coefficients.

f(x) = $(2 + x)\left(1 - \dfrac{5}{2}x + \dfrac{75}{8}x^2 - \dfrac{625}{16}x^3 + \ldots\right)$

$2 \times \dfrac{75}{8} + 1 \times -\dfrac{5}{2} = \dfrac{65}{4}$

x^2 term is $\dfrac{65}{4}x^2$

b The expansion is valid if $|5x| < 1$

$\Rightarrow |x| < \dfrac{1}{5}$

Online Use your calculator to calculate the coefficients of the binomial expansion.

Problem-solving

There are two ways to make an x^2 term.
Either $2 \times \dfrac{75}{8}x^2$ or $x \times \dfrac{5}{2}x$

Add these together to find the term in x^2.

Example **5** **SKILLS** PROBLEM-SOLVING

In the expansion of $(1 + kx)^{-4}$ the coefficient of x is 20.

a Find the value of k.

b Find the corresponding coefficient of the x^2 term.

---

a $(1 + kx)^{-4} = 1 + (-4)(kx) + \dfrac{(-4)(-5)}{2!}(kx)^2 + \ldots$ •————————— Find the binomial expansion of $(1 + kx)^{-4}$

$= 1 - 4kx + 10k^2x^2 + \ldots$

$-4k = 20$
$k = -5$]————————— Solve to find k.

b Coefficient of x^2 = $10k^2$ = $10(-5)^2$ = 250

Exercise (**4A**)   **SKILLS** ▷ PROBLEM-SOLVING

1 For each of the following:

   i find the binomial expansion up to and including the x^3 term

   ii state the range of values of x for which the expansion is valid.

   a $(1 + x)^{-4}$ **b** $(1 + x)^{-6}$ **c** $(1 + x)^{\frac{1}{2}}$

   d $(1 + x)^{\frac{5}{3}}$ **e** $(1 + x)^{-\frac{1}{4}}$ **f** $(1 + x)^{-\frac{3}{2}}$

2 For each of the following:

   i find the binomial expansion up to and including the x^3 term

   ii state the range of values of x for which the expansion is valid.

   a $(1 + 3x)^{-3}$ **b** $\left(1 + \frac{1}{2}x\right)^{-5}$ **c** $(1 + 2x)^{\frac{3}{4}}$

   d $(1 - 5x)^{\frac{7}{3}}$ **e** $(1 + 6x)^{-\frac{2}{3}}$ **f** $\left(1 - \frac{3}{4}x\right)^{-\frac{5}{3}}$

3 For each of the following:

   i find the binomial expansion up to and including the x^3 term

   ii state the range of values of x for which the expansion is valid.

   a $\dfrac{1}{(1 + x)^2}$ **b** $\dfrac{1}{(1 + 3x)^4}$ **c** $\sqrt{1 - x}$

   d $\sqrt[3]{1 - 3x}$ **e** $\dfrac{1}{\sqrt{1 + \frac{1}{2}x}}$ **f** $\dfrac{\sqrt[3]{1 - 2x}}{1 - 2x}$

> **Hint** In part **f**, write the fraction as a single **power** of $(1 - 2x)$

(**E/P**) **4** $f(x) = \dfrac{1 + x}{1 - 2x}$

   a Show that the series expansion of $f(x)$ up to and including the x^3 term is $1 + 3x + 6x^2 + 12x^3$ **(4 marks)**

   b State the range of values of x for which the expansion is valid.

                                                      (1 mark)

> **Hint** First rewrite $f(x)$ as $(1 + x)(1 - 2x)^{-1}$

(**E**) **5** $f(x) = \sqrt{1 + 3x}, \ -\dfrac{1}{3} < x < \dfrac{1}{3}$

   a Find the series expansion of $f(x)$, in **ascending** powers of x, up to and including the x^3 term. Simplify each term.   **(4 marks)**

   b Show that, when $x = \dfrac{1}{100}$, the exact value of $f(x)$ is $\dfrac{\sqrt{103}}{10}$ **(2 marks)**

   c Find the **percentage** error made in using the series expansion in part **a** to **estimate** the value of $f(0.01)$. Give your answer to 2 **significant figures**.   **(3 marks)**

(**P**) **6** In the expansion of $(1 + ax)^{-\frac{1}{2}}$ the coefficient of x^2 is 24.

   a Find the possible values of a.

   b Find the corresponding coefficient of the x^3 term.

(P) **7** Show that if x is small, the expression $\sqrt{\dfrac{1+x}{1-x}}$ is approximated by $1 + x + \dfrac{1}{2}x^2$

Notation 'x is small' means we can assume the expansion is valid for the x values being considered because high powers become **insignificant** compared to the first few terms.

(E/P) **8** $\text{h}(x) = \dfrac{6}{1+5x} - \dfrac{4}{1-3x}$

   a Find the series expansion of $\text{h}(x)$, in ascending powers of x, up to and including the x^2 term. Simplify each term. **(6 marks)**

   b Find the percentage error made in using the series expansion in part **a** to estimate the value of $\text{h}(0.01)$. Give your answer to 2 significant figures. **(3 marks)**

   c Explain why it is not valid to use the expansion to find $\text{h}(0.5)$. **(1 mark)**

(E/P) **9 a** Find the binomial expansion of $(1 - 3x)^{\frac{3}{2}}$ in ascending powers of x up to and including the x^3 term, simplifying each term. **(4 marks)**

   b Show that, when $x = \dfrac{1}{100}$, the exact value of $(1-3x)^{\frac{3}{2}}$ is $\dfrac{97\sqrt{97}}{1000}$ **(2 marks)**

   c Substitute $x = \dfrac{1}{100}$ into the binomial expansion in part **a** and hence obtain an approximation to $\sqrt{97}$. Give your answer to 5 decimal places. **(3 marks)**

Challenge

$\text{h}(x) = \left(1 + \dfrac{1}{x}\right)^{-\frac{1}{2}}, |x| > 1$

a Find the binomial expansion of $\text{h}(x)$ in ascending powers of x up to and including the x^2 term, simplifying each term.

Hint Replace x with $\dfrac{1}{x}$

b Show that, when $x = 9$, the exact value of $\text{h}(x)$ is $\dfrac{3\sqrt{10}}{10}$

c Use the expansion in part **a** to find an approximate value of $\sqrt{10}$. Write your answer to 2 decimal places.

4.2 Expanding $(a + bx)^n$

The binomial expansion of $(1 + x)^n$ can be used to expand $(a + bx)^n$ for any constants a and b.

You need to take a factor of a^n out of the expression:

$$(a + bx)^n = \left(a\left(1 + \dfrac{b}{a}x\right)\right)^n = a^n\left(1 + \dfrac{b}{a}x\right)^n$$

Watch out Make sure you multiply a^n by **every term** in the expansion of $\left(1 + \dfrac{b}{a}x\right)^n$

- The expansion of $(a + bx)^n$, where n is negative or a fraction, is valid for $\left|\dfrac{b}{a}x\right| < 1$ or $|x| < \left|\dfrac{a}{b}\right|$

Example   **6**   **SKILLS**   **ADAPTIVE LEARNING**

Find the first four terms in the binomial expansion of   **a** $\sqrt{4 + x}$ **b** $\dfrac{1}{(2 + 3x)^2}$

State the range of values of x for which each of these expansions is valid.

---

a $\sqrt{4 + x} = (4 + x)^{\frac{1}{2}}$ •—————————   Write in index form.

$$= \left(4\left(1 + \frac{x}{4}\right)\right)^{\frac{1}{2}}$$ Take out a factor of $4^{\frac{1}{2}}$

$$= 4^{\frac{1}{2}}\left(1 + \frac{x}{4}\right)^{\frac{1}{2}}$$ •—————————   Write $4^{\frac{1}{2}}$ as 2.

$$= 2\left(1 + \frac{x}{4}\right)^{\frac{1}{2}}$$

$$= 2\left(1 + \left(\frac{1}{2}\right)\left(\frac{x}{4}\right) + \frac{\left(\frac{1}{2}\right)\left(\frac{1}{2} - 1\right)\left(\frac{x}{4}\right)^2}{2!}\right.$$

Expand $\left(1 + \dfrac{x}{4}\right)^{\frac{1}{2}}$ using the binomial expansion with $n = \dfrac{1}{2}$ and $x = \dfrac{x}{4}$

$$\left. + \frac{\left(\frac{1}{2}\right)\left(\frac{1}{2} - 1\right)\left(\frac{1}{2} - 2\right)\left(\frac{x}{4}\right)^3}{3!} + \dots\right)$$

$$= 2\left(1 + \left(\frac{1}{2}\right)\left(\frac{x}{4}\right) + \frac{\left(\frac{1}{2}\right)\left(-\frac{1}{2}\right)\left(\frac{x^2}{16}\right)}{2}\right.$$

$$\left. + \frac{\left(\frac{1}{2}\right)\left(-\frac{1}{2}\right)\left(-\frac{3}{2}\right)\left(\frac{x^3}{64}\right)}{6} + \dots\right)$$

Simplify coefficients.

$$= 2\left(1 + \frac{x}{8} - \frac{x^2}{128} + \frac{x^3}{1024} + \dots\right)$$

Multiply every term in the expansion by 2.

$$= 2 + \frac{x}{4} - \frac{x^2}{64} + \frac{x^3}{512} + \dots$$

Expansion is valid if $\left|\dfrac{x}{4}\right| < 1$ •—————

The expansion is infinite, and **converges** when $\left|\dfrac{x}{4}\right| < 1$, or $|x| < 4$

$$\Rightarrow |x| < 4$$

b $\dfrac{1}{(2 + 3x)^2} = (2 + 3x)^{-2}$ ●————————————— Write in index form.

$$= \left(2\left(1 + \frac{3x}{2}\right)\right)^{-2}$$

$$= 2^{-2}\left(1 + \frac{3x}{2}\right)^{-2}$$ ————— Take out a factor of 2^{-2}

$$= \frac{1}{4}\left(1 + \frac{3x}{2}\right)^{-2}$$ ————— Write $2^{-2} = \dfrac{1}{2^2} = \dfrac{1}{4}$

$$= \frac{1}{4}\left(1 + (-2)\left(\frac{3x}{2}\right) + \frac{(-2)(-2-1)\left(\frac{3x}{2}\right)^2}{2!}\right.$$

$$\left. + \frac{(-2)(-2-1)(-2-2)\left(\frac{3x}{2}\right)^3}{3!} + \dots\right)$$

Expand $\left(1 + \dfrac{3x}{2}\right)^{-2}$ using the binomial expansion with $n = -2$ and $x = \dfrac{3x}{2}$

$$= \frac{1}{4}\left((1 + (-2)\left(\frac{3x}{2}\right) + \frac{(-2)(-3)\left(\frac{9x^2}{4}\right)}{2}\right.$$

$$\left. + \frac{(-2)(-3)(-4)\left(\frac{27x^3}{8}\right)}{6} + \dots\right)$$

Simplify coefficients.

$$= \frac{1}{4}\left(1 - 3x + \frac{27x^2}{4} - \frac{27x^3}{2} + \dots\right)$$ ————— Multiply every term by $\dfrac{1}{4}$

$$= \frac{1}{4} - \frac{3}{4}x + \frac{27x^2}{16} - \frac{27x^3}{8} + \dots$$

Expansion is valid if $\left|\dfrac{3x}{2}\right| < 1$ ●

$$\Rightarrow |x| < \frac{2}{3}$$

The expansion is infinite, and converges when $\left|\dfrac{3x}{2}\right| < 1$, $|x| < \dfrac{2}{3}$

Exercise (**4B**)   **SKILLS** ▸ ANALYSIS

(P) **1** For each of the following:

  i find the binomial expansion up to and including the x^3 term

  ii state the range of values of x for which the expansion is valid.

  a $\sqrt{4 + 2x}$ **b** $\dfrac{1}{2 + x}$ **c** $\dfrac{1}{(4 - x)^2}$ **d** $\sqrt{9 + x}$

  e $\dfrac{1}{\sqrt{2 + x}}$ **f** $\dfrac{5}{3 + 2x}$ **g** $\dfrac{1 + x}{2 + x}$ **Hint** Write part **g** as $1 - \dfrac{1}{x + 2}$ **h** $\sqrt{\dfrac{2 + x}{1 - x}}$

(E) **2** $f(x) = (5 + 4x)^{-2}$, $|x| < \dfrac{5}{4}$

Find the binomial expansion of $f(x)$ in ascending powers of x, up to and including the term in x^3. Give each coefficient as a simplified fraction. **(5 marks)**

(E) **3** $m(x) = \sqrt{4 - x}$, $|x| < 4$

    a Find the series expansion of $m(x)$, in ascending powers of x, up to and including the x^2 term. Simplify each term. **(4 marks)**

    b Show that, when $x = \dfrac{1}{9}$, the exact value of $m(x)$ is $\dfrac{\sqrt{35}}{3}$ **(2 marks)**

    c Use your answer to part **a** to find an approximate value for $\sqrt{35}$, and calculate the percentage error in your approximation. **(4 marks)**

(P) **4** The first three terms in the binomial expansion of $\dfrac{1}{\sqrt{a + bx}}$ are $3 + \dfrac{1}{3}x + \dfrac{1}{18}x^2 + \ldots$

    a Find the values of the constants a and b.

    b Find the coefficient of the x^3 term in the expansion.

(P) **5** $f(x) = \dfrac{3 + 2x - x^2}{4 - x}$

Prove that if x is **sufficiently small**, $f(x)$ may be approximated by $\dfrac{3}{4} + \dfrac{11}{16}x - \dfrac{5}{64}x^2$

(E/P) **6 a** Expand $\dfrac{1}{\sqrt{5 + 2x}}$, where $|x| < \dfrac{5}{2}$, in ascending powers of x up to and including the term in x^2, giving each coefficient in simplified surd form. **(5 marks)**

    b Hence or otherwise, find the first 3 terms in the expansion of $\dfrac{2x - 1}{\sqrt{5 + 2x}}$ as a series in ascending powers of x. **(4 marks**

(E/P) **7 a** Use the binomial theorem to expand $(16 - 3x)^{\frac{1}{4}}$, $|x| < \dfrac{16}{3}$ in ascending powers of x, up to and including the term in x^2, giving each term as a simplified fraction. **(4 marks)**

    b Use your expansion, with a suitable value of x, to obtain an approximation to $\sqrt[4]{15.7}$ Give your answer to 3 decimal places. **(2 marks)**

8 $g(x) = \dfrac{3}{4 - 2x} - \dfrac{2}{3 + 5x}$, $|x| < \dfrac{1}{2}$

    a Show that the first three terms in the series expansion of $g(x)$ can be written as $\dfrac{1}{12} + \dfrac{107}{72}x - \dfrac{719}{432}x^2$ **(5 marks)**

    b Find the exact value of $g(0.01)$. Round your answer to 7 decimal places. **(2 marks)**

    c Find the percentage error made in using the series expansion in part **a** to estimate the value of $g(0.01)$. Give your answer to 2 significant figures. **(3 marks)**

4.3 Using partial fractions

Partial fractions can be used to simplify the expansions of more difficult expressions.

> **Links**   You need to be confident expressing **algebraic** fractions as sums of partial fractions.

Example **7**   **SKILLS**   INNOVATION

a Express $\dfrac{4 - 5x}{(1 + x)(2 - x)}$ as partial fractions.

b Hence show that the **cubic** approximation of $\dfrac{4 - 5x}{(1 + x)(2 - x)}$ is $2 - \dfrac{7x}{2} + \dfrac{11}{4}x^2 - \dfrac{25}{8}x^3$

c State the range of values of x for which the expansion is valid.

---

a $\dfrac{4 - 5x}{(1 + x)(2 - x)} \equiv \dfrac{A}{1 + x} + \dfrac{B}{2 - x}$ — The denominators must be $(1 + x)$ and $(2 - x)$

$\equiv \dfrac{A(2 - x) + B(1 + x)}{(1 + x)(2 - x)}$ — Add the fractions.

$4 - 5x \equiv A(2 - x) + B(1 + x)$ — Set the numerators equal.

Substitute $x = 2$:

$\quad 4 - 10 = A \times 0 + B \times 3$ — Set $x = 2$ to find B.

$\quad -6 = 3B$

$\quad B = -2$

Substitute $x = -1$:

$\quad 4 + 5 = A \times 3 + B \times 0$ — Set $x = -1$ to find A.

$\quad 9 = 3A$

$\quad A = 3$

so $\dfrac{4 - 5x}{(1 + x)(2 - x)} = \dfrac{3}{1 + x} - \dfrac{2}{2 - x}$

     — Write in index form.

b $\dfrac{4 - 5x}{(1 + x)(2 - x)} = \dfrac{3}{1 + x} - \dfrac{2}{2 - x}$

$= 3(1 + x)^{-1} - 2(2 - x)^{-1}$

> **Problem-solving**
>
> Use headings to keep track of your working. This will help you stay organised and check your answers.

The expansion of $3(1 + x)^{-1}$

$= 3\left(1 + (-1)x + (-1)(-2)\dfrac{x^2}{2!}\right.$

$\left. + (-1)(-2)(-3)\dfrac{x^3}{3!} + ...\right)$

     — Expand $3(1 + x)^{-1}$ using the binomial expansion with $n = -1$

$= 3(1 - x + x^2 - x^3 + ...)$

$= 3 - 3x + 3x^2 - 3x^3 + ...$

The expansion of $2(2 - x)^{-1}$

$$= 2\left(2\left(1 - \frac{x}{2}\right)\right)^{-1}$$

$$= 2 \times 2^{-1}\left(1 - \frac{x}{2}\right)^{-1}$$ Take out a factor of 2^{-1}

$$= 1 \times \left(1 + (-1)\left(-\frac{x}{2}\right) + \frac{(-1)(-2)\left(-\frac{x}{2}\right)^2}{2!}\right.$$
$$\left. + \frac{(-1)(-2)(-3)\left(-\frac{x}{2}\right)^3}{3!} + \dots\right)$$

Expand $\left(1 - \frac{x}{2}\right)^{-1}$ using the binomial expansion with $n = -1$ and $x = \frac{x}{2}$

$$= 1 \times \left(1 + \frac{x}{2} + \frac{x^2}{4} + \frac{x^3}{8} + \dots\right)$$

$$= 1 + \frac{x}{2} + \frac{x^2}{4} + \frac{x^3}{8}$$

Hence $\dfrac{4 - 5x}{(1 + x)(2 - x)}$

$$= 3(1 + x)^{-1} - 2(2 - x)^{-1}$$ 'Add' both expressions.

$$= (3 - 3x + 3x^2 - 3x^3)$$
$$- \left(1 + \frac{x}{2} + \frac{x^2}{4} + \frac{x^3}{8}\right)$$

$$= 2 - \frac{7}{2}x + \frac{11}{4}x^2 - \frac{25}{8}x^3$$ The expansion is infinite, and converges when $|x| < 1$

c $\dfrac{3}{1 + x}$ is valid if $|x| < 1$

$\dfrac{2}{2 - x}$ is valid if $\left|\dfrac{x}{2}\right| < 1 \Rightarrow |x| < 2$ The expansion is infinite, and converges when $\left|\dfrac{x}{2}\right| < 1$ or $|x| < 2$

The expansion is valid when $|x| < 1$

Watch out   You need to find the range of values of x that satisfy **both** inequalities.

Exercise **4C**   **SKILLS**   INNOVATION

(P) **1 a** Express $\dfrac{8x + 4}{(1 - x)(2 + x)}$ as partial fractions.

    b Hence or otherwise expand $\dfrac{8x + 4}{(1 - x)(2 + x)}$ in ascending powers of x as far as the term in x^2.

    c State the set of values of x for which the expansion is valid.

(P) **2** **a** Express $-\dfrac{2x}{(2 + x)^2}$ as partial fractions.

   b Hence prove that $-\dfrac{2x}{(2 + x)^2}$ can be expressed in the form $-\dfrac{1}{2}x + Bx^2 + Cx^3$ where constants B
   and C are to be determined.

   c State the set of values of x for which the expansion is valid.

(P) **3** **a** Express $\dfrac{6 + 7x + 5x^2}{(1 + x)(1 - x)(2 + x)}$ as partial fractions.

   b Hence or otherwise expand $\dfrac{6 + 7x + 5x^2}{(1 + x)(1 - x)(2 + x)}$ in ascending powers of x as far as the term in x^3.

   c State the set of values of x for which the expansion is valid.

(E/P) **4** $g(x) = \dfrac{12x - 1}{(1 + 2x)(1 - 3x)}, |x| < \dfrac{1}{3}$

   Given that $g(x)$ can be expressed in the form $g(x) = \dfrac{A}{1 + 2x} + \dfrac{B}{1 - 3x}$

   a Find the values of A and B. **(3 marks)**

   b Hence, or otherwise, find the series expansion of $g(x)$, in ascending powers of x,
   up to and including the x^2 term. Simplify each term. **(6 marks)**

(P) **5** **a** Express $\dfrac{2x^2 + 7x - 6}{(x + 5)(x - 4)}$ in partial fractions.

> **Hint** First divide the numerator by the denominator.

   b Hence, or otherwise, expand $\dfrac{2x^2 + 7x - 6}{(x + 5)(x - 4)}$ in ascending

   powers of x as far as the term in x^2.

   c State the set of values of x for which the expansion is valid.

(E/P) **6** $\dfrac{3x^2 + 4x - 5}{(x + 3)(x - 2)} = A + \dfrac{B}{x + 3} + \dfrac{C}{x - 2}$

   a Find the values of the constants A, B and C. **(4 marks)**

   b Hence, or otherwise, expand $\dfrac{3x^2 + 4x - 5}{(x + 3)(x - 2)}$ in ascending powers of x, as far as the term in x^2.

   Give each coefficient as a simplified fraction. **(7 marks)**

(E/P) **7** $f(x) = \dfrac{2x^2 + 5x + 11}{(2x - 1)^2(x + 1)}, |x| < \dfrac{1}{2}$

   $f(x)$ can be expressed in the form $f(x) = \dfrac{A}{2x - 1} + \dfrac{B}{(2x - 1)^2} + \dfrac{C}{x + 1}$

   a Find the values of A, B and C. **(4 marks)**

   b Hence or otherwise, find the series expansion of $f(x)$, in ascending powers of x,
   up to and including the term in x^2. Simplify each term. **(6 marks)**

   c Find the percentage error made in using the series expansion in part **b** to estimate
   the value of $f(0.05)$. Give your answer to 2 significant figures. **(4 marks)**

Chapter review 4　　**SKILLS**　PROBLEM-SOLVING

(P) **1** For each of the following

　　i find the binomial expansion up to and including the x^3 term

　　ii state the range of values of x for which the expansion is valid.

　　a $(1 - 4x)^3$　　　　**b** $\sqrt{16 + x}$　　　　**c** $\dfrac{1}{1 - 2x}$　　　　**d** $\dfrac{4}{2 + 3x}$

　　e $\dfrac{4}{\sqrt{4 - x}}$　　　**f** $\dfrac{1 + x}{1 + 3x}$　　　**g** $\left(\dfrac{1 + x}{1 - x}\right)^2$　　　**h** $\dfrac{x - 3}{(1 - x)(1 - 2x)}$

(E) **2** Use the binomial expansion to expand $\left(1 - \dfrac{1}{2}x\right)^{\frac{1}{2}}$, $|x| < 2$ in ascending powers of x, up to and including the term in x^3, simplifying each term.　　**(5 marks)**

3 a Give the binomial expansion of $(1 + x)^{\frac{1}{2}}$ up to and including the term in x^3.

　　b By substituting $x = \dfrac{1}{4}$, find an **approximation** to $\sqrt{5}$ as a fraction.

(E/P) **4** The binomial expansion of $(1 + 9x)^{\frac{2}{3}}$ in ascending powers of x up to and including the term in x^3 is $1 + 6x + cx^2 + dx^3$, $|x| < \dfrac{1}{9}$

　　a Find the value of c and the value of d.　　**(4 marks)**

　　b Use this expansion with your values of c and d together with an appropriate value of x to obtain an estimate of $(1.45)^{\frac{2}{3}}$　　**(2 marks)**

　　c Obtain $(1.45)^{\frac{2}{3}}$ from your calculator and hence make a comment on the accuracy of the estimate you obtained in part **b**.　　**(1 mark)**

(P) **5** In the expansion of $(1 + ax)^{\frac{1}{2}}$ the coefficient of x^2 is -2.

　　a Find the possible values of a.

　　b Find the corresponding coefficients of the x^3 term.

(E) **6** $f(x) = (1 + 3x)^{-1}$, $|x| < \dfrac{1}{3}$

　　a Expand $f(x)$ in ascending powers of x up to and including the term in x^3.　　**(5 marks)**

　　b Hence show that, for small x:

$$\frac{1 + x}{1 + 3x} \approx 1 - 2x + 6x^2 - 18x^3$$　　**(4 marks)**

　　c Taking a suitable value for x, which should be stated, use the series expansion in part **b** to find an approximate value for $\dfrac{101}{103}$, giving your answer to 5 decimal places. **(3 marks)**

(E/P) **7** When $(1 + ax)^n$ is expanded as a series in ascending powers of x, the coefficients of x and x^2 are -6 and 27 respectively.

　　a Find the values of a and n.　　**(4 marks)**

　　b Find the coefficient of x^3.　　**(3 marks)**

　　c State the values of x for which the expansion is valid.　　**(1 mark)**

8 Show that if x is sufficiently small then $\dfrac{3}{\sqrt{4+x}}$ can be approximated by $\dfrac{3}{2} - \dfrac{3}{16}x + \dfrac{9}{256}x^2$

(E) **9 a** Expand $\dfrac{1}{\sqrt{4-x}}$, where $|x| < 4$, in ascending powers of x up to and including the term in x^2.

Simplify each term. **(5 marks)**

b Hence, or otherwise, find the first 3 terms in the expansion of $\dfrac{1+2x}{\sqrt{4-x}}$ as a series in

ascending powers of x. **(4 marks)**

(E) **10 a** Find the first four terms of the expansion, in ascending powers of x, of

$(2 + 3x)^{-1}, |x| < \dfrac{2}{3}$ **(4 marks)**

b Hence or otherwise, find the first four **non-zero** terms of the expansion, in ascending powers of x, of:

$\dfrac{1+x}{2+3x}, |x| < \dfrac{2}{3}$ **(3 marks)**

(E/P) **11 a** Use the binomial theorem to expand $(4 + x)^{-\frac{1}{2}}, |x| < 4$, in ascending powers of x, up to and including the x^3 term, giving each answer as a simplified fraction. **(5 marks)**

b Use your expansion, together with a suitable value of x, to obtain an approximation

to $\dfrac{\sqrt{2}}{2}$. Give your answer to 4 decimal places. **(3 marks)**

(E) **12** $q(x) = (3 + 4x)^{-3}, |x| < \dfrac{3}{4}$

Find the binomial expansion of $q(x)$ in ascending powers of x, up to and including the term in the x^2. Give each coefficient as a simplified fraction. **(5 marks)**

(E/P) **13** $g(x) = \dfrac{39x + 12}{(x + 1)(x + 4)(x - 8)}, |x| < 1$

$g(x)$ can be expressed in the form $g(x) = \dfrac{A}{x + 1} + \dfrac{B}{x + 4} + \dfrac{C}{x - 8}$

a Find the values of A, B and C. **(4 marks)**

b Hence, or otherwise, find the series expansion of $g(x)$, in ascending powers of x, up to and including the x^2 term. Simplify each term. **(7 marks)**

(E/P) **14** $f(x) = \dfrac{12x + 5}{(1 + 4x)^2}, |x| < \dfrac{1}{4}$

For $x \neq -\dfrac{1}{4}, \dfrac{12x + 5}{(1 + 4x)^2} = \dfrac{A}{1 + 4x} + \dfrac{B}{(1 + 4x)^2}$, where A and B are constants.

a Find the values of A and B. **(3 marks)**

b Hence, or otherwise, find the series expansion of $f(x)$, in ascending powers of x, up to and including the term x^2, simplifying each term. **(6 marks)**

(E/P) **15** $q(x) = \dfrac{9x^2 + 26x + 20}{(1 + x)(2 + x)}, |x| < 1$

    a Show that the expansion of $q(x)$ in ascending powers of x can be approximated to $10 - 2x + Bx^2 + Cx^3$ where B and C are constants to be found.   **(7 marks)**

    b Find the percentage error made in using the series expansion in part **a** to estimate the value of $q(0.1)$. Give your answer to 2 significant figures.   **(4 marks)**

Challenge

Obtain the first four non-zero terms in the expansion, in ascending powers of x, of the function $f(x)$ where $f(x) = \dfrac{1}{\sqrt{1 + 3x^2}}, 3x^2 < 1$

Summary of key points

1 This form of the binomial expansion can be applied to negative or fractional values of n to obtain an infinite series:

$$(1 + x)^n = 1 + nx + \frac{n(n - 1)x^2}{2!} + \frac{n(n - 1)(n - 2)x^3}{3!} + \ldots + \frac{n(n - 1)\ldots(n - r + 1)x^r}{r!} + \ldots$$

The expansion is valid when $|x| < 1, n \in \mathbb{R}$.

2 The expansion of $(1 + bx)^n$, where n is negative or a fraction, is valid for $|bx| < 1$, or $|x| < \dfrac{1}{|b|}$

3 The expansion of $(a + bx)^n$, where n is negative or a fraction, is valid for $\left|\dfrac{b}{a}x\right| < 1$ or $|x| < \left|\dfrac{a}{b}\right|$

4 If an expression is of the form $\dfrac{f(x)}{g(x)}$ where $g(x)$ can be split into linear factors, then split $\dfrac{f(x)}{g(x)}$ into partial fractions before expanding each part of the new expression.

Review exercise

E/P **1** Prove by contradiction that there are infinitely many prime numbers. **(4)**

← Pure 4 Section 1.1

E/P **2** Prove that the equation $x^2 - 2 = 0$ has no rational solutions.
You may assume that if n^2 is an even integer then n is also an even integer. **(4)**

← Pure 4 Section 1.1

E **3** Prove by contradiction, that if n is odd, then $3n^2 + 2$ is odd. **(4)**

← Pure 4 Section 1.1

P **4** Prove by contradiction that $\sqrt{5}$ is irrational.

← Pure 4 Section 1.1

E **5** Show that $\dfrac{2x-1}{(x-1)(2x-3)}$ can be written in the form $\dfrac{A}{x-1} + \dfrac{B}{2x-3}$ where A and B are constants to be found. **(3)**

← Pure 4 Section 2.1

E **6** Given that

$$\frac{3x+7}{(x+1)(x+2)(x+3)} \equiv \frac{P}{x+1} + \frac{Q}{x+2} + \frac{R}{x+3}$$

where P, Q and R are constants, find the values of P, Q and R. **(4)**

← Pure 4 Section 2.1

E **7** $f(x) = \dfrac{2}{(2-x)(1+x)^2}, x \neq -1, x \neq 2$

Find the values of A, B and C such that

$$f(x) = \frac{A}{2-x} + \frac{B}{1+x} + \frac{C}{(1+x)^2}$$ **(4)**

← Pure 4 Section 2.2

8 $\dfrac{14x^2 + 13x + 2}{(x+1)(2x+1)^2}$

$$\equiv \frac{A}{x+1} + \frac{B}{2x+1} + \frac{C}{(2x+1)^2}$$

Find the values of the constants A, B and C. ← Pure 4 Section 2.2

E **9** Given that $\dfrac{3x^2 + 6x - 2}{x^2 + 4} \equiv d + \dfrac{ex+f}{x^2+4}$

find the values of d, e and f. **(4)**

← Pure 4 Section 2.3

E **10** $p(x) = \dfrac{9 - 3x - 12x^2}{(1-x)(1+2x)}$

Show that $p(x)$ can be written in the form $A + \dfrac{B}{1-x} + \dfrac{C}{1+2x}$, where A, B and C are constants to be found. **(4)**

← Pure 4 Sections 2.1, 2.3

E **11** Split $\dfrac{4x-1}{(x+1)(x+3)}$ into partial fractions. ← Pure 4 Section 2.1

E/P **12** Given that $\dfrac{4x^3}{(x-3)(x-1)^2}$ can be written as $A + \dfrac{B}{x-3} + \dfrac{C}{x-1} + \dfrac{D}{(x-1)^2}$ determine the values of A, B, C and D.

← Pure 4 Sections 2.1, 2.3

E/P **13 a** Express $\dfrac{5x+3}{(2x-3)(x+2)}$ in partial fractions. **(3)**

b Hence find the exact value of

$$\int_2^6 \frac{5x+3}{(2x-3)(x+2)} \, dx, \text{ giving your}$$

answer as a single logarithm. **(4)**

← Pure 4 Section 2.1

E **14** A curve C has parametric equations

$$x = 1 - \frac{4}{t}, y = t^2 - 3t + 1, t \in R, t \neq 0$$

a Determine the ranges of x and y in the given domain of t. **(3)**

b Show that the Cartesian equation of C can be written in the form

$$y = \frac{ax^2 + bx + c}{(1 - x)^2}, \text{ where } a, b \text{ and } c \text{ are}$$

integers to be found. **(3)**

← Pure 4 Section 3.1

E **15** A curve has parametric equations

$$x = \ln(t + 2), y = \frac{3t}{t + 3}, t > 4$$

a Find a Cartesian equation of this curve in the form $y = f(x)$, $x > k$, where k is an exact constant to be found. **(4)**

b Write down the range of $f(x)$ in the form $a < y < b$, where a and b are constants to be found. **(2)**

← Pure 4 Section 3.1

E **16** A curve C has parametric equations

$$x = \frac{1}{1 + t}, y = \frac{1}{1 - t}, -1 < t < 1$$

Show that a Cartesian equation of C is

$$y = \frac{x}{2x - 1}$$ **(4)**

← Pure 4 Section 3.1

E/P **17** A curve C has parametric equations

$$x = 2 \cos t, y = \cos 3t, 0 \leqslant t \leqslant \frac{\pi}{2}$$

a Find a Cartesian equation of the curve in the form $y = f(x)$, where $f(x)$ is a cubic function. **(5)**

b State the domain and range of $f(x)$ for the given domain of t. **(2)**

← Pure 4 Section 3.2

E/P **18** The curve shown in the figure has parametric equations

$$x = \sin t, y = \sin\left(t + \frac{\pi}{6}\right), -\frac{\pi}{2} \leqslant t \leqslant \frac{\pi}{2}$$

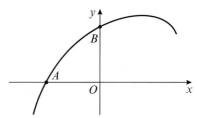

a Show that a Cartesian equation of the curve is

$$y = \frac{\sqrt{3}}{2}x + \frac{1}{2}\sqrt{(1 - x^2)}, -1 \leqslant x \leqslant 1$$ **(4)**

b Find the coordinates of the points A and B, where the curve intercepts the x- and y-axes. **(3)**

← Pure 4 Section 3.2

E **19** The curve C has parametric equations

$$x = 3 \cos t, y = \cos 2t, 0 \leqslant t \leqslant \pi$$

a Find a Cartesian equation of C. **(4)**

b Sketch the curve C on the appropriate domain, labelling the points where the curve intercepts the x- and y-axes. **(3)**

← Pure 4 Section 3.2, 3.3

E **20** **a** Expand $(1 - 2x)^{10}$ in ascending powers of x up to and including the term in x^3. **(3)**

b Use your answer to part **a** to evaluate $(0.98)^{10}$ correct to 3 decimal places. **(1)**

← Pure 4 Section 4.1

E/P **21** If x is so small that terms of x^3 and higher can be ignored,

$$(2 - x)(1 + 2x)^5 \approx a + bx + cx^2$$

Find the values of the constants a, b and c. **(5)**

← Pure 4 Section 4.1

E/P **22** The coefficient of x in the binomial expansion of $(2 - 4x)^q$, where q is a positive integer, is $-32q$. Find the value of q. **(4)**

← Pure 4 Section 4.2

(E) 23 $g(x) = \dfrac{1}{\sqrt{1-x}}$

 a Show that the series expansion of $g(x)$ up to and including the x^3 term is
$$1 + \frac{x}{2} + \frac{3x^2}{8} + \frac{5x^3}{16} \qquad \textbf{(5)}$$

 b State the range of values of x for which the expansion is valid. **(1)**

← Pure 4 Section 4.1

(P) 24 When $(1 + ax)^n$ is expanded as a series in ascending powers of x, the coefficients of x and x^2 are -6 and 45 respectively.

 a Find the value of a and the value of n.

 b Find the coefficient of x^3.

 c Find the set of values of x for which the expansion is valid.

← Pure 4 Section 4.1

(E) 25 a Find the binomial expansion of $(1 + 4x)^{\frac{3}{2}}$ in ascending powers of x up to and including the x^3 term, simplifying each term. **(4)**

 b Show that, when $x = \dfrac{3}{100}$, the exact

value of $(1 + 4x)^{\frac{3}{2}}$ is $\dfrac{112\sqrt{112}}{1000}$ **(2)**

 c Substitute $x = \dfrac{3}{100}$ into the binomial expansion in part **a** and hence obtain an approximation to $\sqrt{112}$. Give your answer to 5 decimal places. **(3)**

 d Calculate the percentage error in your estimate to 5 decimal places. **(2)**

← Pure 4 Section 4.1

(E) 26 $f(x) = (1 + x)(3 + 2x)^{-3}$, $|x| < \dfrac{3}{2}$

 Find the binomial expansion of $f(x)$ in ascending powers of x, up to and including the term in x^3. Give each coefficient as a simplified fraction. **(5)**

← Pure 4 Section 4.2

(E) 27 $h(x) = \sqrt{4 - 9x}$, $|x| < \dfrac{4}{9}$

 a Find the series expansion of $h(x)$, in ascending powers of x, up to and including the x^2 term. Simplify each term. **(4)**

 b Show that, when $x = \dfrac{1}{100}$, the exact

value of $h(x)$ is $\dfrac{\sqrt{391}}{10}$ **(2)**

 c Use the series expansion in part **a** to estimate the value of $h\left(\dfrac{1}{100}\right)$ and state the degree of accuracy of your approximation. **(3)**

← Pure 4 Section 4.2

(E/P) 28 Given that $(a + bx)^{-2}$ has binomial

expansion $\dfrac{1}{4} + \dfrac{1}{4}x + cx^2 + \ldots$

 a Find the values of the constants a, b and c. **(4)**

 b Find the coefficient of the x^3 term in the expansion. **(2)**

← Pure 4 Section 4.2

(E/P) 29 $g(x) = \dfrac{3 + 5x}{(1 + 3x)(1 - x)}$, $|x| < \dfrac{1}{3}$

 Given that $g(x)$ can be expressed in the form $g(x) = \dfrac{A}{1 + 3x} + \dfrac{B}{1 - x}$

 a find the values of A and B. **(3)**

 b Hence, or otherwise, find the series expansion of $f(x)$, in ascending powers of x, up to and including the x^2 term. Simplify each term. **(6)**

← Pure 4 Sections 4.1, 4.3

(E/P) 30 $\dfrac{3x - 1}{(1 - 2x)^2} \equiv \dfrac{A}{1 - 2x} + \dfrac{B}{(1 - 2x)^2}$, $|x| < \dfrac{1}{2}$

 a Find the values of A and B. **(3)**

 b Hence, or otherwise, expand $\dfrac{3x - 1}{(1 - 2x)^2}$ in ascending powers of x, as far as the term in x^3. Give each coefficient as a simplified fraction. **(6)**

← Pure 4 Sections 4.1, 4.3

(E/P) **31** $f(x) = \dfrac{25}{(3 + 2x)^2(1 - x)}$, $|x| < 1$

f(x) can be expressed in the form

$$\frac{A}{3 + 2x} + \frac{B}{(3 + 2x)^2} + \frac{C}{1 - x}$$

a Find the values of A, B and C. **(4)**

b Hence, or otherwise, find the series expansion of f(x), in ascending powers of x, up to and including the term in x^2. Simplify each term. **(6)**

← **Pure 4 Sections 4.1, 4.2, 4.3**

(E/P) **32** $\dfrac{4x^2 + 30x + 31}{(x + 4)(2x + 3)} = A + \dfrac{B}{x + 4} + \dfrac{C}{2x + 3}$

a Find the values of the constants A, B and C. **(4)**

b Hence, or otherwise, expand $\dfrac{4x^2 + 31x + 30}{(x + 4)(2x + 3)}$ in ascending powers of x, as far as the term in x^2. Give each coefficient as a simplified fraction. **(7)**

← **Pure 4 Sections 4.1, 4.2, 4.3**

Challenge

1 Prove by contradiction that if $a, b \in \mathbb{Z}$ then $a^2 - 8b \neq 2$

2 Express $\dfrac{2x^4 + 3}{x^2 - 1}$ as partial fractions.

3 A curve has parametric equations

$$x = \sin t, \, y = \sin 3t, \, 0 \leqslant t \leqslant \pi$$

a Find a Cartesian equation of the curve in the form $y = f(x)$

b Determine the domain and range of f(x), and hence sketch the curve.

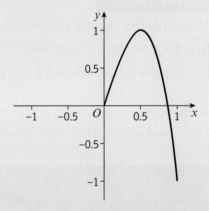

4 Using the binomial expansion, find an approximation, to 6 decimal places, for $\sqrt{98}$ Using your result, find an approximate value for $\cos\dfrac{\pi}{4}$ to 6 decimal places.

5 DIFFERENTIATION

5.1
5.2

Learning objectives

After completing this chapter you should be able to:
* Differentiate parametric equations   → pages 51–52
* Differentiate functions which are defined implicitly
                                                              → pages 54–55
* Solve problems involving connected rates of change
  and construct simple differential equations
                                                              → pages 57–59

Prior knowledge check

1 Differentiate:

   a $\sin 2x$ **b** $x^2(1 - 3x)^4$ **c** $e^{2x + 3}$
                           ← Pure 3 Sections 6.1, 6.2, 6.4

2 Find the gradient of the function $y = \ln\left(\dfrac{2x}{x + 3}\right)$ at the
   point where $x = 1$
                           ← Pure 3 Sections 6.2, 6.5

3 The curve C is defined by the parametric equations
   $$x = 3t^2 - 5t, \quad y = t^3 + 2, \quad t \in \mathbb{R}$$
   Find the coordinates of any points where C intersects
   the coordinate axes.   ← Pure 1 Sections 4.1, 4.3

4 Solve $2\operatorname{cosec} x - 3\sec x = 0$ in the interval $0 \leqslant x \leqslant 2\pi$
   giving your answers correct to 3 significant figures.
                           ← Pure 3 Sections 3.2, 3.4

You can use differentiation to find rates of change in trigonometric and exponential models. The velocity of a wrecking ball could be estimated by modelling its displacement then differentiating.

5.1   Parametric differentiation

When functions are defined parametrically, you can find the **gradient** at a given point without converting into Cartesian form. You can use a variation of the chain rule:

■ If x and y are given as functions of a parameter, t: $\dfrac{dy}{dx} = \dfrac{\dfrac{dy}{dt}}{\dfrac{dx}{dt}}$

> **Hint**   You can obtain this from writing $\dfrac{dy}{dx} \times \dfrac{dx}{dt} = \dfrac{dy}{dt}$

Example 1   SKILLS   PROBLEM-SOLVING

Find the gradient at the point P where $t = 2$, on the curve given parametrically by

$$x = t^3 + t, \quad y = t^2 + 1, \quad t \in \mathbb{R}$$

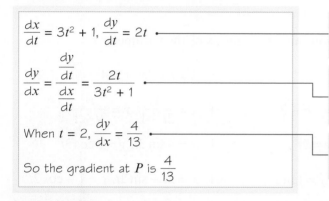

$$\frac{dx}{dt} = 3t^2 + 1, \frac{dy}{dt} = 2t$$

$$\frac{dy}{dx} = \frac{\dfrac{dy}{dt}}{\dfrac{dx}{dt}} = \frac{2t}{3t^2 + 1}$$

When $t = 2$, $\dfrac{dy}{dx} = \dfrac{4}{13}$

So the gradient at P is $\dfrac{4}{13}$

First **differentiate** x and y with respect to the parameter t.

This rule will give the gradient function $\dfrac{dy}{dx}$ in terms of the **parameter**, t.

Substitute $t = 2$ into $\dfrac{2t}{3t^2 + 1}$

Example 2   SKILLS   INNOVATION

Find the equation of the **normal** at the point P where $\theta = \dfrac{\pi}{6}$, to the curve with parametric equations $x = 3 \sin \theta$, $y = 5 \cos \theta$.

$$\frac{dx}{d\theta} = 3 \cos \theta, \frac{dy}{d\theta} = -5 \sin \theta \quad 0 \leqslant \theta < 2\pi$$

$$\therefore \frac{dy}{dx} = \frac{-5 \sin \theta}{3 \cos \theta}$$

At point P, where $\theta = \dfrac{\pi}{6}$

$$\frac{dy}{dx} = \frac{-5 \times \dfrac{1}{2}}{3 \times \dfrac{\sqrt{3}}{2}} = \frac{-5}{3\sqrt{3}}$$

First differentiate x and y with respect to the parameter θ.

Use the chain rule, $\dfrac{dy}{d\theta} \div \dfrac{dx}{d\theta}$, and substitute $\theta = \dfrac{\pi}{6}$

> **Online**   Explore the graph of this curve and the normal at this point using technology.

The gradient of the normal at P is $\dfrac{3\sqrt{3}}{5}$

and at P, $x = \dfrac{3}{2}$, $y = \dfrac{5\sqrt{3}}{2}$

The equation of the normal is

$$y - \frac{5\sqrt{3}}{2} = \frac{3\sqrt{3}}{5}\left(x - \frac{3}{2}\right)$$

$\therefore\ 5y = 3\sqrt{3}x + 8\sqrt{3}$

The normal is **perpendicular** to the curve, so its gradient is $-\dfrac{1}{m}$ where m is the gradient of the curve at that point.

You need to find the coordinates of P. Substitute $\theta = \dfrac{\pi}{6}$ into each of the parametric equations.
← **Pure 2 Section 6.2**

Use the equation for a line in the form
$y - y_1 = m(x - x_1)$

Exercise (5A)　　**SKILLS**　PROBLEM-SOLVING

1 Find $\dfrac{dy}{dx}$ for each of the following, leaving your answer in terms of the parameter t.

a $x = 2t$, $y = t^2 - 3t + 2$　　**b** $x = 3t^2$, $y = 2t^3$　　**c** $x = t + 3t^2$, $y = 4t$

d $x = t^2 - 2$, $y = 3t^5$　　**e** $x = \dfrac{2}{t}$, $y = 3t^2 - 2$　　**f** $x = \dfrac{1}{2t - 1}$, $y = \dfrac{t^2}{2t - 1}$

g $x = \dfrac{2t}{1 + t^2}$, $y = \dfrac{1 - t^2}{1 + t^2}$　　**h** $x = t^2 e^t$, $y = 2t$　　**i** $x = 4\sin 3t$, $y = 3\cos 3t$

j $x = 2 + \sin t$, $y = 3 - 4\cos t$　　**k** $x = \sec t$, $y = \tan t$　　**l** $x = 2t - \sin 2t$, $y = 1 - \cos 2t$

m $x = e^t - 5$, $y = \ln t$, $t > 0$　　**n** $x = \ln t$, $y = t^2 - 64$, $t > 0$　　**o** $x = e^{2t} + 1$, $y = 2e^t - 1$, $-1 < t < 1$

(P) **2 a** Find the equation of the **tangent** to the curve with parametric equations
$x = 3 - 2\sin t$, $y = t\cos t$, at the point P, where $t = \pi$

b Find the equation of the tangent to the curve with parametric equations
$x = 9 - t^2$, $y = t^2 + 6t$, at the point P, where $t = 2$

(P) **3 a** Find the equation of the normal to the curve with parametric equations
$x = e^t$, $y = e^t + e^{-t}$, at the point P, where $t = 0$

b Find the equation of the normal to the curve with parametric equations
$x = 1 - \cos 2t$, $y = \sin 2t$, at the point P, where $t = \dfrac{\pi}{6}$

(P) **4** Find the points of zero gradient on the curve with parametric equations

$$x = \frac{t}{1 - t}, \quad y = \frac{t^2}{1 - t}, \quad t \neq 1$$

You do not need to establish whether they are maximum or minimum points.

(P) **5** The curve C has parametric equations $x = e^{2t}$, $y = e^t - 1$, $t \in \mathbb{R}$

a Find the equation of the tangent to C at the point A where $t = \ln 2$

b Show that the curve C has no **stationary points**.

E/P **6** The curve C has parametric equations

$$x = \frac{t^2 - 3t - 4}{t}, \quad y = 2t, \quad t > 0$$

The line l_1 is a tangent to C and is **parallel** to the line with equation $y = x + 5$
Find the equation of l_1. **(8 marks)**

E/P **7** A curve has parametric equations

$$x = 2\sin^2 t, \quad y = 2\cot t, \quad 0 < t < \frac{\pi}{2}$$

a Find an expression for $\dfrac{dy}{dx}$ in terms of the parameter t. **(4 marks)**

b Find an equation of the tangent to the curve at the point where $t = \dfrac{\pi}{6}$ **(4 marks)**

E/P **8** The curve C has parametric equations

$$x = 4\sin t, \quad y = 2\operatorname{cosec} 2t, \quad 0 \leqslant t \leqslant \pi$$

The point A lies on C and has coordinates $\left(2\sqrt{3}, \dfrac{4\sqrt{3}}{3}\right)$

a Find the value of t at the point A. **(2 marks)**

The line l is a normal to C at A.

b Show that an equation for l is $9x + 12y - 34\sqrt{3} = 0$ **(6 marks)**

E/P **9** The curve C has parametric equations

$$x = t^2 + t, \quad y = t^2 - 10t + 5, \quad t \in \mathbb{R}$$

where t is a parameter. Given that at point P, the gradient of C is 2,

a find the coordinates of P **(4 marks)**

b find the equation of the tangent to C at point P **(3 marks)**

c show that the tangent to C at point P does not intersect the curve again. **(5 marks)**

> **Problem-solving**
>
> Substitute the equations for x and y into the equation of your tangent, and show that the resulting quadratic equation has no real roots.

E/P **10** The curve C has parametric equations

$$x = 2\sin t, \quad y = \sqrt{2}\cos 2t, \quad 0 < t < \pi$$

a Find an expression for $\dfrac{dy}{dx}$ in terms of t. **(2 marks)**

The point A lies on C where $t = \dfrac{\pi}{3}$ The line l is the normal to C at A.

b Find an equation for l in the form $ax + by + c = 0$, where a, b and c are exact constants to be found. **(5 marks)**

c Prove that the line l does not intersect the curve anywhere other than at point A. **(6 marks)**

(E/P) **11** A curve has parametric equations

$$x = \cos t, \quad y = \tfrac{1}{2}\sin 2t, \quad 0 \leqslant t < 2\pi$$

a Find an expression for $\dfrac{dy}{dx}$ in terms of t. **(2 marks)**

b Find an equation of the tangent to the curve at point A where $t = \dfrac{\pi}{6}$ **(4 marks)**

The lines l_1 and l_2 are two further distinct tangents to the curve. Given that l_1 and l_2 are both parallel to the tangent to the curve at point A,

c find an equation of l_1 and an equation of l_2 **(6 marks)**

5.2 Implicit differentiation

Some equations are difficult to rearrange into the form $y = f(x)$ or $x = f(y)$. You can sometimes differentiate these equations **implicitly** without rearranging them.

In general, from the chain rule:

Notation An equation in the form $y = f(x)$ is given **explicitly**.

Equations which involve functions of both x and y such as $x^2 + 2xy = 3$ or $\cos(x + y) = 2x$ are called **implicit** equations.

- $\dfrac{d}{dx}(f(y)) = f'(y)\dfrac{dy}{dx}$

The following two specific results are useful for implicit **differentiation**:

- $\dfrac{d}{dx}(y^n) = ny^{n-1}\dfrac{dy}{dx}$

- $\dfrac{d}{dx}(xy) = x\dfrac{dy}{dx} + y$

When you differentiate implicit equations your expression for $\dfrac{dy}{dx}$ will usually be given in terms of **both x and y**.

Watch out You need to pay careful attention to the variable you are differentiating with respect to.

Example 3 **SKILLS** ADAPTIVE LEARNING

Find $\dfrac{dy}{dx}$ in terms of x and y where $x^3 + x + y^3 + 3y = 6$

$$3x^2 + 1 + 3y^2\dfrac{dy}{dx} + 3\dfrac{dy}{dx} = 0$$

Differentiate the expression term by term with respect to x.

Use $\dfrac{d}{dx}(y^n) = ny^{n-1}\dfrac{dy}{dx}$ with $n = 3$

$$\dfrac{dy}{dx}(3y^2 + 3) = -3x^2 - 1$$

Then make $\dfrac{dy}{dx}$ the subject of the formula.

$$\dfrac{dy}{dx} = -\dfrac{3x^2 + 1}{3(1 + y^2)}$$

Divide both sides by $3y^2 + 3$ and factorise.

Example **4**   **SKILLS**   PROBLEM-SOLVING

Given that $4xy^2 + \dfrac{6x^2}{y} = 10$, find the value of $\dfrac{dy}{dx}$ at the point (1, 1).

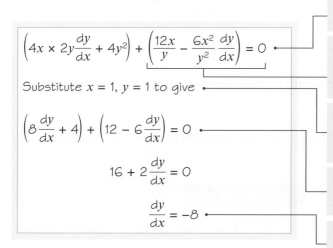

$$\left(4x \times 2y\frac{dy}{dx} + 4y^2\right) + \left(\frac{12x}{y} - \frac{6x^2}{y^2}\frac{dy}{dx}\right) = 0$$

Substitute $x = 1$, $y = 1$ to give

$$\left(8\frac{dy}{dx} + 4\right) + \left(12 - 6\frac{dy}{dx}\right) = 0$$

$$16 + 2\frac{dy}{dx} = 0$$

$$\frac{dy}{dx} = -8$$

Differentiate each term with respect to x.

Use the product rule on each term, expressing $\dfrac{6x^2}{y}$ as $6x^2y^{-1}$

Find the value of $\dfrac{dy}{dx}$ at (1, 1) by substituting $x = 1, y = 1$

Substitute before rearranging, as this simplifies the working.

Solve to find the value of $\dfrac{dy}{dx}$ at this point.

Example **5**   **SKILLS**   PROBLEM-SOLVING

Find the value of $\dfrac{dy}{dx}$ at the point (1, 1) where $e^{2x} \ln y = x + y - 2$.

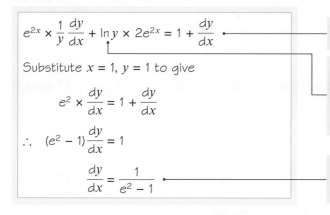

$$e^{2x} \times \frac{1}{y}\frac{dy}{dx} + \ln y \times 2e^{2x} = 1 + \frac{dy}{dx}$$

Substitute $x = 1$, $y = 1$ to give

$$e^2 \times \frac{dy}{dx} = 1 + \frac{dy}{dx}$$

$$\therefore \quad (e^2 - 1)\frac{dy}{dx} = 1$$

$$\frac{dy}{dx} = \frac{1}{e^2 - 1}$$

Differentiate each term with respect to x.

Use the product rule applied to the term on the left hand side of the equation, noting that $\ln y$ differentiates to give $\dfrac{1}{y}\dfrac{dy}{dx}$

Rearrange to make $\dfrac{dy}{dx}$ the subject of the formula.

Exercise **5B**   **SKILLS**   INNOVATION

(P)   **1**   By writing $u = y^n$, and using the chain rule, show that $\dfrac{d}{dx}(y^n) = ny^{n-1}\dfrac{dy}{dx}$

(P)   **2**   Use the product rule to show that $\dfrac{d}{dx}(xy) = x\dfrac{dy}{dx} + y$

(P) **3** Find an expression in terms of x and y for $\dfrac{dy}{dx}$, given that:

     a $x^2 + y^3 = 2$ **b** $x^2 + 5y^2 = 14$ **c** $x^2 + 6x - 8y + 5y^2 = 13$

     d $y^3 + 3x^2y - 4x = 0$ **e** $3y^2 - 2y + 2xy = x^3$ **f** $x = \dfrac{2y}{x^2 - y}$

     g $(x - y)^4 = x + y + 5$ **h** $e^x y = xe^y$ **i** $\sqrt{xy} + x + y^2 = 0$

(P) **4** Find the equation of the tangent to the curve with implicit equation $x^2 + 3xy^2 - y^3 = 9$ at the point $(2, 1)$.

(P) **5** Find the equation of the normal to the curve with implicit equation $(x + y)^3 = x^2 + y$ at the point $(1, 0)$.

(P) **6** Find the coordinates of the points of zero gradient on the curve with implicit equation $x^2 + 4y^2 - 6x - 16y + 21 = 0$

> **Problem-solving**
>
> Find $\dfrac{dy}{dx}$ then set the numerator equal to 0 to find the x-coordinate at the points of 0 gradient. You need to find two corresponding y-coordinates.

(E/P) **7** A curve C is described by the equation

$$2x^2 + 3y^2 - x + 6xy + 5 = 0$$

Find an equation of the tangent to C at the point $(1, -2)$, giving your answer in the form $ax + by + c = 0$, where a, b and c are integers. **(7 marks)**

(E/P) **8** A curve C has equation

$$3^x = y - 2xy$$

Find the exact value of $\dfrac{dy}{dx}$ at the point on C with coordinates $(2, -3)$ **(7 marks)**

(E/P) **9** Find the gradient of the curve with equation

$$\ln(y^2) = \frac{1}{2}x \ln(x - 1), \quad x > 1, \quad y > 0$$

at the point on the curve where $x = 4$. Give your answer as an exact value. **(7 marks)**

(E/P) **10** A curve C satisfies $\sin x + \cos y = 0.5$, where $-\pi < x < \pi$ and $-\pi < y < \pi$

     a Find an expression for $\dfrac{dy}{dx}$ **(2 marks)**

     b Find the coordinates of the stationary points on C. **(5 marks)**

(E/P) **11** The curve C has the equation $ye^{-3x} - 3x = y^2$

     a Find $\dfrac{dy}{dx}$ in terms of x and y. **(5 marks)**

     b Show that the equation of the tangent to C at the **origin**, O, is $y = 3x$ **(4 marks)**

Challenge

The curve C has implicit equation $6x + y^2 + 2xy = x^2$

a Show that there are no points on the curve such that $\dfrac{dy}{dx} = 0$

b Find the coordinates of the two points on C such that $\dfrac{dx}{dy} = 0$

5.3   Rates of change

- You can use the chain rule to connect rates of change in situations involving more than two variables.

Example (6)   **SKILLS**   CRITICAL THINKING

Given that the area of a circle A cm^2 is related to its radius r cm by the formula $A = \pi r^2$, and that the **rate of change** of its radius in cm s^{-1} is given by $\dfrac{dr}{dt} = 5$, find $\dfrac{dA}{dt}$ when $r = 3$

$$A = \pi r^2$$

$$\therefore \quad \frac{dA}{dr} = 2\pi r$$

Using $\dfrac{dA}{dt} = \dfrac{dA}{dr} \times \dfrac{dr}{dt}$

$$\frac{dA}{dt} = 2\pi r \times 5$$

$$= 30\pi, \text{ when } r = 3$$

Problem-solving

In order to be able to apply the chain rule to find $\dfrac{dA}{dt}$ you need to know $\dfrac{dA}{dr}$. You can find it by differentiating $A = \pi r^2$ with respect to r.

You should use the chain rule, giving the **derivative** which you need to find in terms of known derivatives.

Example (7)   **SKILLS**   INTERPRETATION

The volume of a **hemisphere** V cm^3 is related to its radius r cm by the formula $V = \dfrac{2}{3}\pi r^3$ and the total **surface area** S cm^2 is given by the formula $S = \pi r^2 + 2\pi r^2 = 3\pi r^2$. Given that the rate of increase of volume, in cm^3 s^{-1}, $\dfrac{dV}{dt} = 6$, find the rate of increase of surface area $\dfrac{dS}{dt}$

$$V = \frac{2}{3}\pi r^3 \text{ and } S = 3\pi r^2$$

$$\frac{dV}{dr} = 2\pi r^2 \text{ and } \frac{dS}{dr} = 6\pi r$$

This is area of circular base plus area of curved surface.

As V and S are functions of r, find $\dfrac{dV}{dr}$ and $\dfrac{dS}{dr}$

Now $\dfrac{dS}{dt} = \dfrac{dS}{dr} \times \dfrac{dr}{dV} \times \dfrac{dV}{dt}$

$\qquad = 6\pi r \times \dfrac{1}{2\pi r^2} \times 6$

$\qquad = \dfrac{18}{r}$

Use the chain rule together with the property that $\dfrac{dr}{dV} = 1 \div \dfrac{dV}{dr}$

An equation that involves a rate of change is called a **differential equation**. You can formulate differential equations from information given in a question.

Links You can use integration to solve differential equations. → **Pure 4 Section 6.5**

Example 8 **SKILLS** ADAPTIVE LEARNING

In the decay of radioactive particles, the rate at which particles decay is **proportional** to the number of particles remaining. Write down a differential equation for the rate of change of the number of particles.

Let N be the number of particles and let t be time. The rate of change of the number of particles $\dfrac{dN}{dt}$ is proportional to N.

i.e. $\dfrac{dN}{dt} = -kN$, where k is a positive constant.

The minus sign arises because the number of particles is decreasing.

$\dfrac{dN}{dt} \propto N$ so you can write $\dfrac{dN}{dt} = kN$

where k is the constant of proportion.

Example 9 **SKILLS** ADAPTIVE LEARNING

Newton's law of cooling states that the rate of loss of temperature of a body is proportional to the excess temperature of the body over its surroundings. Write an equation that expresses this law.

Let the temperature of the body be θ degrees and the time be t seconds.

The rate of change of the temperature $\dfrac{d\theta}{dt}$ is proportional to $\theta - \theta_0$, where θ_0 is the temperature of the surroundings.

i.e. $\dfrac{d\theta}{dt} = -k(\theta - \theta_0)$, where k is a positive constant.

$\theta - \theta_0$ is the difference between the temperature of the body and that of its surroundings.

The minus sign arises because the temperature is decreasing. The question mentions loss of temperature.

Example **10**　　**SKILLS**　CRITICAL THINKING

The head of a snowman of radius R cm loses volume by evaporation at a rate proportional to its surface area. Assuming that the head is spherical, that the volume of a **sphere** is $\frac{4}{3}\pi R^3$ cm^3 and that the surface is $4\pi R^2$ cm^2, write down a **differential equation** for the rate of change of radius of the snowman's head.

The first sentence tells you that $\dfrac{dV}{dt} = -kA$, where V cm^3 is the volume, t seconds is time, k is a positive constant and A cm^2 is the surface area of the snowman's head.

Since　　$V = \dfrac{4}{3}\pi R^3$

$\dfrac{dV}{dR} = 4\pi R^2$

$\therefore$　　$\dfrac{dV}{dt} = \dfrac{dV}{dR} \times \dfrac{dR}{dt} = 4\pi R^2 \times \dfrac{dR}{dt}$

But as　$\dfrac{dV}{dt} = -kA$

$4\pi R^2 \times \dfrac{dR}{dt} = -k \times 4\pi R^2$

$\therefore$　　$\dfrac{dR}{dt} = -k$

The question asks for a differential equation in terms of R, so you need to use the expression for V in terms of R.

The chain rule is used here because this is a related rate of change.

Use the expression for A in terms of R.

Divide both sides by the common factor $4\pi R^2$.

This gives the rate of change of radius as required.

Exercise **5C**　　**SKILLS**　CREATIVITY

(P)　**1**　Given that $A = \dfrac{1}{4}\pi r^2$ and that $\dfrac{dr}{dt} = 6$, find $\dfrac{dA}{dt}$ when $r = 2$

(P)　**2**　Given that $y = xe^x$ and that $\dfrac{dx}{dt} = 5$, find $\dfrac{dy}{dt}$ when $x = 2$

(P)　**3**　Given that $r = 1 + 3\cos\theta$ and that $\dfrac{d\theta}{dt} = 3$, find $\dfrac{dr}{dt}$ when $\theta = \dfrac{\pi}{6}$

(P)　**4**　Given that $V = \dfrac{1}{3}\pi r^3$ and that $\dfrac{dV}{dt} = 8$, find $\dfrac{dr}{dt}$ when $r = 3$

(P)　**5**　A population is growing at a rate which is proportional to the size of the population. Write down a differential equation for the growth of the population.

(P)　**6**　A curve C has equation $y = f(x)$, $y > 0$. At any point P on the curve, the gradient of C is proportional to the product of the x- and the y-coordinates of P. The point A with coordinates $(4, 2)$ is on C and the gradient of C at A is $\dfrac{1}{2}$

Show that $\dfrac{dy}{dx} = \dfrac{xy}{16}$

(P) **7** Liquid is pouring into a container at a constant rate of $30\,\text{cm}^3\,\text{s}^{-1}$. At time t seconds liquid is leaking from the container at a rate of $\frac{2}{15}V\,\text{cm}^3\,\text{s}^{-1}$, where $V\,\text{cm}^3$ is the volume of the liquid in the container at that time.

Show that $-15\dfrac{\mathrm{d}V}{\mathrm{d}t} = 2V - 450$

(P) **8** An electrically-charged body loses its charge, Q coulombs, at a rate, measured in coulombs per second, proportional to the charge Q.
Write down a differential equation in terms of Q and t where t is the time in seconds since the body started to lose its charge.

(P) **9** The ice on a pond has a thickness x mm at a time t hours after the start of freezing. The rate of increase of x is **inversely proportional** to the square of x.
Write down a differential equation in terms of x and t.

(P) **10** The radius of a circle is increasing at a constant rate of 0.4 cm per second.

a Find $\dfrac{\mathrm{d}C}{\mathrm{d}t}$, where C is the circumference of the circle, and interpret this value in the context of the **model**.

b Find the rate at which the area of the circle is increasing when the radius is 10 cm.

c Find the radius of the circle when its area is increasing at the rate of $20\,\text{cm}^2$ per second.

(P) **11** The volume of a **cube** is decreasing at a constant rate of $4.5\,\text{cm}^3$ per second. Find

a the rate at which the length of one side of the cube is decreasing when the volume is $100\,\text{cm}^3$

b the volume of the cube when the length of one side is decreasing at the rate of 2 mm per second.

(P) **12** Fluid flows out of a **cylindrical** tank with constant cross section. At time t minutes, $t > 0$, the volume of fluid remaining in the tank is $V\,\text{m}^3$. The rate at which the fluid flows in $\text{m}^3\,\text{min}^{-1}$ is proportional to the square root of V.
Show that the depth, h metres, of fluid in the tank satisfies the differential equation $\dfrac{\mathrm{d}h}{\mathrm{d}t} = -k\sqrt{h}$ where k is a positive constant.

(P) **13** At time, t seconds, the surface area of a cube is $A\,\text{cm}^2$ and the volume is $V\,\text{cm}^3$.
The surface area of the cube is expanding at a constant rate of $2\,\text{cm}^2\,\text{s}^{-1}$.

a Write an expression for V in terms of A.

b Find an expression for $\dfrac{\mathrm{d}V}{\mathrm{d}A}$

c Show that $\dfrac{\mathrm{d}V}{\mathrm{d}t} = \dfrac{1}{2}V^{\frac{1}{3}}$

(P) **14** An inverted **conical** funnel is full of salt. The salt is allowed to leave by a small hole in the vertex. It leaves at a constant rate of $6\,\text{cm}^3\,\text{s}^{-1}$.
Given that the angle of the cone between the slanting edge and the vertical is 30°, show that the volume of the salt is $\frac{1}{9}\pi h^3$, where h is the height of salt at time t seconds. Show that the rate of change of the height of the salt in the funnel is inversely proportional to h^2. Write down a differential equation relating h and t.

Chapter review **5** **SKILLS** **PROBLEM-SOLVING**

(E/P) **1** The curve C is given by the equations

$$x = 4t - 3, \quad y = \frac{8}{t^2}, \quad t > 0$$

where t is a parameter.

At A, $t = 2$. The line l is the normal to C at A.

a Find $\dfrac{dy}{dx}$ in terms of t. **(4 marks)**

b Hence find an equation of l. **(3 marks)**

(E/P) **2** The curve C is given by the equations $x = 2t$, $y = t^2$, where t is a parameter.
Find an equation of the normal to C at the point P on C where $t = 3$ **(7 marks)**

(E/P) **3** The curve C has parametric equations

$$x = t^3, \quad y = t^2, \quad t > 0$$

Find an equation of the tangent to C at A $(1, 1)$. **(7 marks)**

(E/P) **4** A curve C is given by the equations

$$x = 2\cos t + \sin 2t, \quad y = \cos t - 2\sin 2t, \quad 0 < t < \pi$$

where t is a parameter.

a Find $\dfrac{dx}{dt}$ and $\dfrac{dy}{dt}$ in terms of t. **(3 marks)**

b Find the value of $\dfrac{dy}{dx}$ at the point P on C where $t = \dfrac{\pi}{4}$ **(3 marks)**

c Find an equation of the normal to the curve at P. **(3 marks)**

(E/P) **5** A curve is given by $x = 2t + 3$, $y = t^3 - 4t$, where t is a parameter. The point A has parameter $t = -1$ and the line l is the tangent to C at A. The line l also cuts the curve at B.

a Show that an equation for l is $2y + x = 7$ **(6 marks)**

b Find the value of t at B. **(5 marks)**

(P) **6** A car has value £V at time t years. A model for V assumes that the rate of decrease of V at time t is proportional to V. Form an appropriate differential equation for V.

(P) **7** In a study of the water loss of picked leaves the mass, M grams, of a single leaf was measured at times, t days, after the leaf was picked. It was found that the rate of loss of mass was proportional to the mass M of the leaf.
Write down a differential equation for the rate of change of mass of the leaf.

(P) **8** In a pond the amount of pondweed, P, grows at a rate proportional to the amount of pondweed already present in the pond. Pondweed is also removed by fish eating it at a constant rate of Q per unit of time.

Write down a differential equation relating P to t, where t is the time which has elapsed since the start of the observation.

(P) **9** A circular patch of oil on the surface of some water has radius r and the radius increases over time at a rate inversely proportional to the radius.
Write down a differential equation relating r and t, where t is the time which has elapsed since the start of the observation.

(P) **10** A metal bar is heated to a certain temperature, then allowed to cool down and it is noted that, at time t, the rate of loss of temperature is proportional to the difference between the temperature of the metal bar, θ, and the temperature of its surroundings θ_0.
Write down a differential equation relating θ and t.

(E/P) **11** The curve C has parametric equations

$$x = 4\cos 2t, \quad y = 3\sin t, \quad -\frac{\pi}{2} < t < \frac{\pi}{2}$$

A is the point $\left(2, \frac{3}{2}\right)$, and lies on C.

a Find the value of t at the point A. **(2 marks)**

b Find $\dfrac{dy}{dx}$ in terms of t. **(3 marks)**

c Show that an equation of the normal to C at A is $6y - 16x + 23 = 0$ **(4 marks)**

The normal at A cuts C again at the point B.

d Find the y-coordinate of the point B. **(6 marks)**

(E/P) **12** The diagram shows the curve C with parametric equations

$$x = a\sin^2 t, \quad y = a\cos t, \quad 0 \le t \le \frac{1}{2}\pi$$

where a is a positive constant. The point P lies on C and has coordinates $\left(\frac{3}{4}a, \frac{1}{2}a\right)$.

a Find $\dfrac{dy}{dx}$, giving your answer in terms of t. **(4 marks)**

b Find an equation of the tangent to C at P. **(4 marks)**

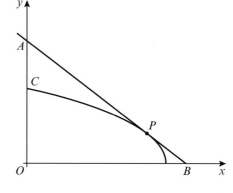

The tangent to C at P cuts the coordinate axes at points A and B.

c Show that the **triangle AOB** has area ka^2 where k is a constant to be found. **(2 marks)**

(E/P) **13** This graph shows part of the curve C with parametric equations

$$x = (t + 1)^2, \quad y = \frac{1}{2}t^3 + 3, \quad t > -1$$

P is the point on the curve where $t = 2$
The line l is the normal to C at P.
Find the equation of l. **(7 marks)**

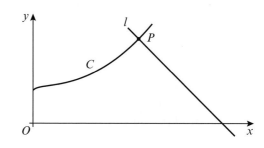

E/P **14** Find the gradient of the curve with equation $5x^2 + 5y^2 - 6xy = 13$ at the point $(1, 2)$. **(7 marks)**

E/P **15** Given that $e^{2x} + e^{2y} = xy$, find $\dfrac{dy}{dx}$ in terms of x and y. **(7 marks)**

E/P **16** Find the coordinates of the turning points on the curve $y^3 + 3xy^2 - x^3 = 3$ **(7 marks)**

E/P **17** **a** If $(1 + x)(2 + y) = x^2 + y^2$, find $\dfrac{dy}{dx}$ in terms of x and y. **(4 marks)**

   b Find the gradient of the curve $(1 + x)(2 + y) = x^2 + y^2$ at each of the two points where the curve meets the y-axis. **(3 marks)**

   c Show also that there are two points at which the tangents to this curve are parallel to the y-axis. **(4 marks)**

E/P **18** A curve has equation $7x^2 + 48xy - 7y^2 + 75 = 0$. A and B are two distinct points on the curve and at each of these points the gradient of the curve is equal to $\dfrac{2}{11}$. Use implicit differentiation to show that the straight line passing through A and B has equation $x + 2y = 0$ **(6 marks)**

E/P **19** Given that $y = x^x$, $x > 0$, $y > 0$, by taking **logarithms** show that

$$\frac{dy}{dx} = x^x(1 + \ln x)$$ **(6 marks)**

E/P **20** **a** Given that $a^x \equiv e^{kx}$, where a and k are constants, $a > 0$ and $x \in \mathbb{R}$, prove that $k = \ln a$. **(2 marks)**

   b Hence, using the derivative of e^{kx}, prove that when $y = 2^x$

$$\frac{dy}{dx} = 2^x \ln 2$$ **(4 marks)**

   c Hence **deduce** that the gradient of the curve with equation $y = 2^x$ at the point $(2, 4)$ is $\ln 16$. **(3 marks)**

E/P **21** A population P is growing at the rate of 9% each year and at time t years may be approximated by the formula

$$P = P_0(1.09)^t, \ t \geqslant 0$$

where P is regarded as a continuous function of t and P_0 is the population at time $t = 0$.

   a Find an expression for t in terms of P and P_0. **(2 marks)**

   b Find the time T years when the population has doubled from its value at $t = 0$, giving your answer to 3 significant figures. **(4 marks)**

   c Find, as a multiple of P_0, the rate of change of population $\dfrac{dP}{dt}$ at time $t = T$ **(4 marks)**

E/P **22** A curve C has equation

$$y = \ln(\sin x), \quad 0 < x < \pi$$

   a Find the stationary point of the curve C. **(6 marks)**

   b Show that the curve C is **concave** at all values of x in its given domain. **(3 marks)**

23 The mass of a radioactive substance t years after first being observed is **modelled** by the equation

$$m = 40e^{-0.244t}$$

a Find the mass of the substance nine months after it was first observed. **(2 marks)**

b Find $\dfrac{dm}{dt}$ **(2 marks)**

c With reference to the model, interpret the significance of the sign of the value of $\dfrac{dm}{dt}$ found in part **b**. **(1 mark)**

24 The curve C with equation $y = f(x)$ is shown in the diagram, where

$$f(x) = \frac{\cos 2x}{e^x}, \ 0 \leqslant x \leqslant \pi$$

The curve has a local minimum at A and a local maximum at B.

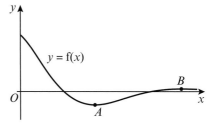

a Show that the x-coordinates of A and B satisfy the equation $\tan 2x = -0.5$ and hence find the coordinates of A and B. **(6 marks)**

b Using your answer to part **a**, find the coordinates of the maximum and minimum turning points on the curve with equation $y = 2 + 4f(x - 4)$ **(3 marks)**

c Determine the range of values for which $f(x)$ is concave. **(5 marks)**

Challenge

The curve C has parametric equations

$$y = 2 \sin 2t, \quad x = 5 \cos\left(t + \frac{\pi}{12}\right), \quad 0 \leqslant t \leqslant 2\pi$$

a Find $\dfrac{dy}{dx}$ in terms of t.

b Find the coordinates of the points on C where $\dfrac{dy}{dx} = 0$

c Find the coordinates of any points where the curve cuts or intersects the coordinate axes, and determine the gradient of the curve at these points.

d Find the coordinates of the points on C where $\dfrac{dx}{dy} = 0$

e Hence sketch C.

Hint The points on C where $\dfrac{dx}{dy} = 0$ correspond to points where a tangent to the curve would be a vertical line.

Summary of key points

1 If x and y are given as functions of a parameter, t: $\dfrac{\mathrm{d}y}{\mathrm{d}x} = \dfrac{\frac{\mathrm{d}y}{\mathrm{d}t}}{\frac{\mathrm{d}x}{\mathrm{d}t}}$

2 For implicit differentiation, the following techniques can be applied.

- $\dfrac{\mathrm{d}}{\mathrm{d}x}(\mathrm{f}(y)) = \mathrm{f}'(y)\dfrac{\mathrm{d}y}{\mathrm{d}x}$

- $\dfrac{\mathrm{d}}{\mathrm{d}x}(y^n) = ny^{n-1}\dfrac{\mathrm{d}y}{\mathrm{d}x}$

- $\dfrac{\mathrm{d}}{\mathrm{d}x}(xy) = x\dfrac{\mathrm{d}y}{\mathrm{d}x} + y$

3 You can use the chain rule to connect rates of change in situations involving more than two variables.

6 INTEGRATION

Learning objectives

After completing this chapter you should be able to:

* Find the area under a curve represented in parametric form
* Find the volume of revolution when a curve is rotated around the x-axis.

→ pages 67–73

* Integrate functions by making a substitution, using integration by parts and using partial fractions → pages 74–83
* Solve simple differential equations and model real-life situations with differential equations → pages 84–91

Prior knowledge check

1 Differentiate:

   a $(2x - 7)^6$ **b** $\sin 5x$ **c** $e^{\frac{x}{3}}$

   ← Pure 3 Sections 5.1, 5.2, 5.3

2 Given $f(x) = 8x^{\frac{1}{2}} - 6x^{-\frac{1}{2}}$ find:

   a $\int f(x)\,dx$

   b $\int_4^9 f(x)\,dx$ ← Pure 1 Section 9.1
   ← Pure 2 Section 8.1

3 Write $\dfrac{3x + 22}{(4x - 1)(x + 3)}$ as partial fractions. ← Pure 4 Section 2.1

4 Find the area of the region R bounded by the curve $y = x^2 + 1$, the x-axis and the lines $x = -1$ and $x = 2$

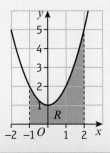

← Pure 2 Section 8.2

Integration can be used to solve differential equations. Archaeologists use differential equations to estimate the age of fossilised plants and animals.

6.1 Finding the area under a curve defined parametrically

You already know how to find the area under a curve using integration. We are now going to extend this idea by looking at the area under a curve that is defined by a set of parametric equations.

Links ← Pure 2 Section 8.2

If we consider the curve C defined by $x = f(t)$, $y = g(t)$ for $a \le t \le b$, and start with the formula for

the area under a curve, then we have $A = \int_{x_1}^{x_2} y\, dx$. The curve we have does not quite fit this

formula. We can see that y can be replaced by $g(t)$, but how to we replace the dx?

Since we know $x = f(t)$, it is reasonable to state that $\dfrac{dx}{dt} = f'(t)$, and so $dx = f'(t)dt$

Now we can write $A = \int_{t=a}^{t=b} g(t)f'(t)dt$. This is the formula for the area under a parametric curve.

Example 1

A curve is represented by the parametric equations $x = t$, $y = 3t^2$
Find the area under the curve from $t = 1$ to $t = 2$

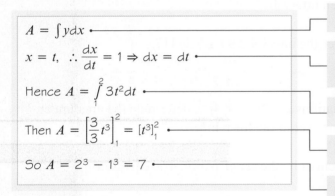

$A = \int y\,dx$	Start with the formula for area.
$x = t, \ \therefore \dfrac{dx}{dt} = 1 \Rightarrow dx = dt$	Differentiate the function x, and obtain dx in terms of dt.
Hence $A = \int_1^2 3t^2 dt$	Re-write the area formula, now in terms of t.
Then $A = \left[\dfrac{3}{3}t^3\right]_1^2 = [t^3]_1^2$	Integrate and simplify.
So $A = 2^3 - 1^3 = 7$	Enter the limits to obtain the answer.

Example 2

A curve is represented by the parametric equations $x = \cos t$, $y = -\cos t$
Find the area under the curve from $t = \dfrac{\pi}{6}$ to $t = \dfrac{\pi}{4}$

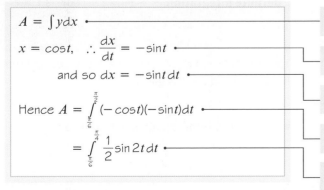

$A = \int y\,dx$	Start with the formula for area.
$x = \cos t, \ \therefore \dfrac{dx}{dt} = -\sin t$	Differentiate the function x.
and so $dx = -\sin t\,dt$	Determine an expression for dx.
Hence $A = \int_{\frac{\pi}{6}}^{\frac{\pi}{2}} (-\cos t)(-\sin t)dt$	Substitute into the area formula.
$= \int_{\frac{\pi}{6}}^{\frac{\pi}{4}} \dfrac{1}{2}\sin 2t\,dt$	Use $\sin 2A = 2\sin A \cos A$

So $A = \left[-\dfrac{1}{4} \cos 2t \right]_{\frac{\pi}{6}}^{\frac{\pi}{4}}$ •————————— Integrate the function.

Then $A = -\dfrac{1}{4}\left(\cos\dfrac{\pi}{2} - \cos\dfrac{\pi}{3} \right)$ •————————— Substitute in the limits.

Which gives $A = -\dfrac{1}{4}\left(-\dfrac{1}{2} \right) = \dfrac{1}{8}$ •————————— Evaluate the area.

Exercise **6A** **SKILLS** ▶ **ANALYSIS**

1 Determine the area under the curve represented by the set of parametric equations
$x = t^2$, $y = t + 1$ for $-1 \le t \le 2$

2 The parametric curve C has the set of equations $x = \sqrt{t}$, $y = 4\,t^{\frac{3}{2}}$

Determine the area under the curve from $t = 3$ to $t = 5$

3 The parametric equations $x = \sin t$, $y = 2\cos t$ represent the curve C, over the interval $0 \le t \le \dfrac{\pi}{3}$

Find the area under the curve over the given interval.

4 The area under the curve $x = 6t^2$, $y = \dfrac{t}{4}$, over the interval $2 \le t \le b$, is 117.

Determine the value of b.

5 Given that the curve C has parametric equations $x = 2t^{\frac{1}{2}}$, $y = 4t^{-\frac{1}{2}}$, determine the area under
the curve from $t = 4$ to $t = 16$, giving your answer in the form $\ln b$.

6 Determine the area under the curve represented by the set of parametric equations
$x = e^{2t}$, $y = e^{3t}$ for $\ln 3 \le t \le \ln 8$

Give your answer in an exact form.

6.2 Volumes of revolution around the *x*-axis

You have used integration to find the area of a
region R bounded by a curve, the x-axis and
two vertical lines.

Notation This process is called
definite integration. ← **Pure 2 Sections 8.1, 8.2**

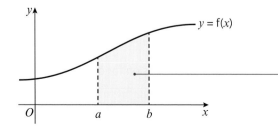

The area between a positive curve, the x-axis and the
lines $x = a$ and $x = b$ is given by
$$\text{Area} = \int_a^b y \, dx$$
where $y = f(x)$ is the equation of the curve.

You can derive this formula by considering the sum of an infinite number of small **strips** of width δx. Each of these strips has a height of y, so the area of each strip is

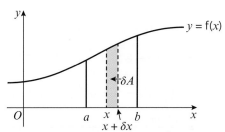

$$\delta A = y\delta x$$

The total area is approximately the sum of these strips, or $\sum y\delta x$

The exact area is the **limit** of this sum as $\delta x \to 0$ which is written as $\int y \, dx$.

You can use a similar technique to find the volume of an object created by rotating a curve around a coordinate axis. If each of these strips is rotated through 2π radians (or $360°$) about the x-axis, it will form a shape that is approximately cylindrical. The volume of each cylinder will be $\pi y^2 \delta x$ since it will have radius y and height δx.

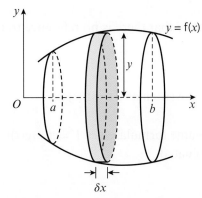

So the volume of the solid will be approximately equal to the sum of the volumes of each cylinder, or $\sum \pi y^2 \delta x$. The exact volume is the limit of this sum as $\delta x \to 0$, or $\pi \int y^2 \, dx$

- **The volume of revolution formed when $y = f(x)$ is rotated through 2π radians about the x-axis between $x = a$ and $x = b$ is given by:**

$$\textbf{Volume} = \pi \int_a^b y^2 \, \textbf{d}x$$

Example 3   SKILLS   PROBLEM-SOLVING

The diagram shows the region R which is bounded by the x-axis, the y-axis and the curve with equation $y = 9 - x^2$

The region is rotated through 2π radians about the x-axis. Find the exact volume of the solid **generated**.

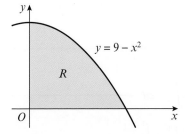

$$9 - x^2 = 0$$

$$(3 - x)(3 + x) = 0$$

$$\underline{x = 3} \text{ or } x = -3$$

First find the point where the curve intersects the x-axis.

From the diagram, $x > 0$, therefore $x = 3$

$$V = \pi \int_0^3 (9 - x^2)^2 \, dx$$

Use $V = \pi \int_a^b y^2 \, dx$ with $a = 0$, $b = 3$ and $y = 9 - x^2$

$$= \pi \int_0^3 (81 - 18x^2 + x^4) \, dx$$

Simplify the **integrand**.

$$= \pi \left[81x - 6x^3 + \frac{1}{5}x^5 \right]_0^3$$

Integrate each term separately.

$$= \pi \left(\left(243 - 162 + \frac{243}{5} \right) - (0 - 0 + 0) \right)$$

Substitute the limits.

$$= \frac{648\pi}{5}$$

Simplify the resulting answer and write it as an exact fraction in terms of π.

If a function is given in parametric form, such as $x = f(t)$, $y = g(t)$ for $a \leq t \leq b$, in order to determine the volume you must start from the standard formula.

So, using $V = \pi \int_{x_1}^{x_2} y^2 \, dx$, we first notice that there is no reference to our parameter, the integral is purely in terms of x.

If we consider $\dfrac{dx}{dt} = f'(t)$, this can be written as $dx = f'(t) \, dt$.

Substituting all of this into our volume formula gives $V = \pi \int_2^b (g(t))^2 \, f'(t) dt$ that is an integral purely in terms of t which can now be evaluated.

Example 4 **SKILLS** PROBLEM-SOLVING

A curve is represented by the parametric equations $x = 2t$, $y = t^2$

Find the volume generated when the curve is rotated about the x-axis, from $t = 2$ to $t = 4$

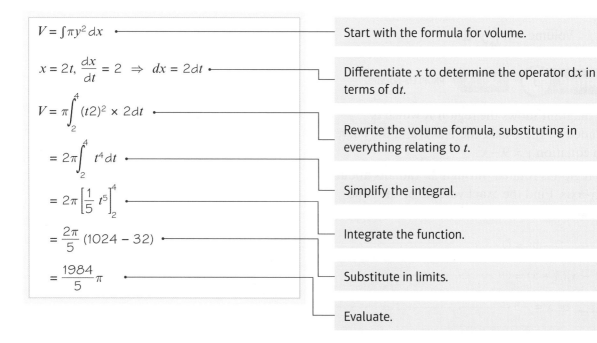

$$V = \int \pi y^2 \, dx$$

Start with the formula for volume.

$$x = 2t, \frac{dx}{dt} = 2 \Rightarrow dx = 2dt$$

Differentiate x to determine the operator dx in terms of dt.

$$V = \pi \int_2^4 (t2)^2 \times 2dt$$

Rewrite the volume formula, substituting in everything relating to t.

$$= 2\pi \int_2^4 t^4 \, dt$$

Simplify the integral.

$$= 2\pi \left[\frac{1}{5} t^5 \right]_2^4$$

Integrate the function.

$$= \frac{2\pi}{5} (1024 - 32)$$

Substitute in limits.

$$= \frac{1984}{5} \pi$$

Evaluate.

Example 5   **SKILLS**   PROBLEM-SOLVING

The curve C has parametric equations $x = 4t^2 - 1$, $y = t^2 - t$, the curve cuts the x-axis at the points A and B. the curve is rotated about the x-axis, between the points A and B. Find the volume generated.

$y = 0,\ t = 0,\ 1$	Determine where the curve cuts the x-axis.
$V = \int \pi y^2\ dx$	Use the formula for volume.
$x = 4t^2 - 1,\ \dfrac{dx}{dt} = 8t\ \Rightarrow\ dx = 8t\,dt$	Differentiate x to determine the operator dx in terms of dt.
$V = \pi\displaystyle\int_0^1 (t^2 - t)^2 \times 8t\ dt$	Rewrite the volume formula, substituting in everything relating to t.
$= 8\pi\displaystyle\int_0^1 (t^4 - 2t^3 + t^2)t\ dt$	
$= 8\pi\displaystyle\int_0^1 (t^5 - 2t^4 + t^3)\ dt$	Simplify the integral.
$= 8\pi\left[\dfrac{1}{6}t^6 - \dfrac{2}{5}t^5 + \dfrac{1}{4}t^4\right]_0^1$	Integrate the function.
$= 8\pi\left(\dfrac{1}{6} - \dfrac{2}{5} + \dfrac{1}{4}\right)$	Substitute in limits.
$= \dfrac{2}{15}\pi$	Evaluate.

Exercise 6B   **SKILLS**   ANALYSIS

1 Find the exact volume of the solid generated when each curve is rotated through 360° about the x-axis between the given limits.

   a $y = 10x^2$ between $x = 0$ and $x = 2$

   b $y = 5 - x$ between $x = 3$ and $x = 5$

   c $y = \sqrt{x}$ between $x = 2$ and $x = 10$

   d $y = 1 + \dfrac{1}{x^2}$ between $x = 1$ and $x = 2$

(E)   **2**   The curve shown in the diagram has equation $y = 5 + 4x - x^2$. The finite region R is bounded by the curve, the x-axis and the y-axis. The region is rotated through 2π radians about the x-axis to generate a solid of revolution. Find the exact volume of the solid generated.   **(5 marks)**

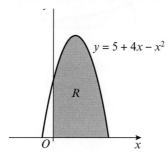

(E) **3** The diagram shows the region R which is bounded by the x-axis, the lines $x = 1$ and $x = 8$, and the curve with equation $y = 3 - \sqrt[3]{x}$. The region is rotated through 2π radians about the x-axis. Find the exact volume of the solid generated. **(5 marks)**

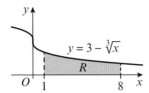

(E) **4** The diagram shows the curve C with equation $y = \sqrt{x + 2}$. The region R is bounded by the x-axis, the line $x = 2$ and C. The region is rotated through $360°$ about the x-axis. Find the exact volume of the solid generated. **(5 marks)**

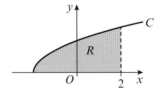

(E/P) **5** The diagram shows a sketch of the curve with equation $y = 9x^{\frac{3}{2}} - 3x^{\frac{5}{2}}$
The region R is bounded by the curve and the x-axis.

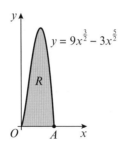

a Find the coordinates of A. **(2 marks)**

The region is rotated through 2π radians about the x-axis.

b Find the volume of the solid of revolution generated. **(5 marks)**

(E/P) **6** The curve with equation $y = \dfrac{\sqrt{3x^4 - 3}}{x^3}$ is shown in the diagram.

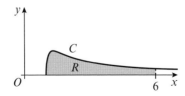

The region bounded by the curve C, the x-axis and the line $x = 6$ is shown shaded in the diagram. The region is rotated through 2π radians about the x-axis. Find the volume of the solid generated, giving your answer correct to 3 significant figures. **(6 marks)**

(P) 7 The diagram shows the curve with equation $5y^2 - x^3 = 2x - 3$. The shaded region is bounded by the curve and the line $x = 4$. The region is rotated about the x-axis to generate a solid of revolution. Find the volume of the solid generated.

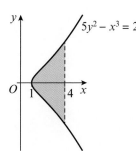

$5y^2 - x^3 = 2x - 3$

Hint

Rearrange the equation to make y^2 the subject.

(E/P) 8 The curve shown in the diagram has equation $y = x\sqrt{4 - x^2}$. The finite region R is bounded by the curve, the x-axis and the line $x = a$, where $0 < a < 2$. The region is rotated through 2π radians about the x-axis to generate a solid of revolution with volume $\dfrac{657\pi}{160}$

Find the value of a. **(5 marks)**

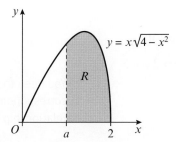

$y = x\sqrt{4 - x^2}$

R

(P) 9 The diagram shows a shaded rectangular region R of length h and width r. The region R is rotated through 360° about the x-axis. Use integration to show that the volume, V, of the cylinder formed is $V = \pi r^2 h$

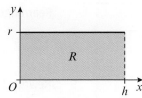

R

(P) 10 A curve is represented by the parametric equations $x = t^{\frac{3}{2}}, y = t^{\frac{1}{2}}$

The curve is rotated about the x-axis to form a volume from $t = 1$ to $t = 3$

Find the volume generated.

(P) 11 The parametric curve C is represented by the set of equations $x = t^2 + 1, y = \dfrac{3}{t}$

The curve is then rotated about the x-axis between the values of $t = 2$ and $t = 3$

Determine the volume generated, giving your answer in the form $k\pi \ln\left(\dfrac{a}{b}\right)$

Challenge

The diagram shows the curve C with equation $y = |x^2 - 7x + 10|$. The shaded region R is bounded by the x-axis, the curve C and the lines $x = 1$ and $x = 6$. The region is rotated 2π radians about the x-axis. Find the exact volume of the solid generated.

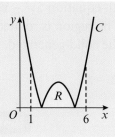

6.3 Integration by substitution

Sometimes you can simplify an **integral** by changing the variable. The process is similar to using the chain rule in differentiation and is called integration by substitution.

In your exam you will often be told which substitution to use.

Example 6   **SKILLS**   ADAPTIVE LEARNING

Find $\int x\sqrt{2x + 5}\, dx$ using the substitutions

a $u = 2x + 5$ **b** $u^2 = 2x + 5$

a Let $I = \int x\sqrt{2x + 5}\, dx$

Let $u = 2x + 5$

So $\dfrac{du}{dx} = 2$

So dx can be replaced by $\dfrac{1}{2}\, du$

$\sqrt{2x + 5} = \sqrt{u} = u^{\frac{1}{2}}$

$x = \dfrac{u - 5}{2}$

So $I = \int \left(\dfrac{u - 5}{2}\right)u^{\frac{1}{2}} \times \dfrac{1}{2}\, du$

$= \int \dfrac{1}{4}(u - 5)u^{\frac{1}{2}}\, du$

$= \int \dfrac{1}{4}(u^{\frac{3}{2}} - 5u^{\frac{1}{2}})\, du$

$= \dfrac{1}{4} \times \dfrac{u^{\frac{5}{2}}}{\frac{5}{2}} - \dfrac{5u^{\frac{3}{2}}}{4 \times \frac{3}{2}} + c$

$= \dfrac{u^{\frac{5}{2}}}{10} - \dfrac{5u^{\frac{3}{2}}}{6} + c$

So $I = \dfrac{(2x + 5)^{\frac{5}{2}}}{10} - \dfrac{5(2x + 5)^{\frac{3}{2}}}{6} + c$

You need to replace each 'x' term with a corresponding 'u' term. Start by finding the relationship between dx and du.

So $dx = \dfrac{1}{2}\, du$

Next rewrite the function in terms of $u = 2x + 5$

Rearrange $u = 2x + 5$ to get $2x = u - 5$ and hence $x = \dfrac{u - 5}{2}$

Rewrite I in terms of u and simplify.

Multiply out the brackets and integrate.
← **Pure 1 Section 9.1**

Simplify.

Finally rewrite the answer in terms of x.

b Let $\quad I = \int x\sqrt{2x+5}\,dx$

$\qquad u^2 = 2x + 5$

$\qquad 2u\dfrac{du}{dx} = 2$

First find the relationship between dx and du.

Using implicit differentiation, cancel 2 and rearrange to get $dx = u\,du$

So replace dx with $u\,du$.

$\sqrt{2x + 5} = u$

and $\quad x = \dfrac{u^2 - 5}{2}$

Rewrite the integrand in terms of u. You will need to make x the subject of $u^2 = 2x + 5$

So $\quad I = \int\left(\dfrac{u^2 - 5}{2}\right)u \times u\,du$

$\qquad = \int\left(\dfrac{u^4}{2} - \dfrac{5u^2}{2}\right)du$

Multiply out the brackets and integrate.

$\qquad = \dfrac{u^5}{10} - \dfrac{5u^3}{6} + c$

So $\quad I = \dfrac{(2x + 5)^{\frac{5}{2}}}{10} - \dfrac{5(2x + 5)^{\frac{3}{2}}}{6} + c$

Rewrite answer in terms of x.

Example **7**   **SKILLS**   **CRITICAL THINKING**

Use the substitution $u = \sin x + 1$ to find

$$\int \cos x \sin x\,(1 + \sin x)^3\,dx$$

Let $\quad I = \int \cos x \sin x\,(1 + \sin x)^3\,dx$

Let $\quad u = \sin x + 1$

$\qquad \dfrac{du}{dx} = \cos x$

First replace the dx.

$\cos x$ appears in the integrand, so you can write this as $du = \cos x\,dx$ and substitute.

So substitute $\cos x\,dx$ with du.

$(\sin x + 1)^3 = u^3$

$\sin x = u - 1$

Use $u = \sin x + 1$ to substitute for the remaining terms, rearranging where required to get $\sin x = u - 1$

So $\quad I = \int (u - 1)u^3\,du$

Rewrite I in terms of u.

$\qquad = \int (u^4 - u^3)\,du$

$\qquad = \dfrac{u^5}{5} - \dfrac{u^4}{4} + c$

Multiply out the brackets and integrate in the usual way.

So $\quad I = \dfrac{(\sin x + 1)^5}{5} - \dfrac{(\sin x + 1)^4}{4} + c$

Problem-solving

Although it looks different, $\int \sin 2x\,(1 + \sin x)^3\,dx$ can be integrated in exactly the same way. Remember $\sin 2x \equiv 2\sin x \cos x$, so the above integral would just need adjusting by a factor of 2.

Example 8   SKILLS   CREATIVITY

Use integration by substitution to **evaluate**

a $\displaystyle\int_0^2 x(x+1)^3\,dx$ **b** $\displaystyle\int_0^{\frac{\pi}{2}} \cos x\sqrt{1+\sin x}\,dx$

Watch out If you use integration by substitution to evaluate a definite integral, you have to be careful of whether your limits are x values or u values. You can use a table to keep track.

a Let $I = \displaystyle\int_0^2 x(x+1)^3\,dx$

   Let $u = x + 1$

   $\dfrac{du}{dx} = 1$

   so replace dx with du and replace $(x+1)^3$ with u^3, and x with $u-1$

Replace each term in x with a term in u in the usual way.

x	u
2	3
0	1

Change the limits.
When $x = 2$, $u = 2 + 1 = 3$ and when $x = 0$, $u = 1$

   So $I = \displaystyle\int_1^3 (u-1)u^3\,du$

Note that the new u limits replace their corresponding x limits.

   $= \displaystyle\int_1^3 (u^4 - u^3)\,du$

   $= \left[\dfrac{u^5}{5} - \dfrac{u^4}{4}\right]_1^3$

Multiply out and integrate. Remember there is no need for a constant of integration with definite integrals.

   $= \left(\dfrac{243}{5} - \dfrac{81}{4}\right) - \left(\dfrac{1}{5} - \dfrac{1}{4}\right)$

   $= 48.4 - 20 = 28.4$

The integral can now be evaluated using the limits for u without having to change back into x.

b $\displaystyle\int_0^{\frac{\pi}{2}} \cos x\sqrt{1+\sin x}\,dx$

   $u = 1 + \sin x \Rightarrow \dfrac{du}{dx} = \cos x$, so replace $\cos x\,dx$ with du and replace $\sqrt{1+\sin x}$ with $u^{\frac{1}{2}}$.

Use $u = 1 + \sin x$

x	u
$\dfrac{\pi}{2}$	2
0	1

Remember that limits for integrals involving trigonometric functions will always be in radians. $x = \dfrac{\pi}{2}$, means $u = 1 + 1 = 2$ and $x = 0$, means $u = 1 + 0 = 1$

   So $I = \displaystyle\int_1^2 u^{\frac{1}{2}}\,du$

Rewrite the integral in terms of u.

   $= \left[\dfrac{2}{3}u^{\frac{3}{2}}\right]_1^2$

   $= \left(\dfrac{2}{3}2^{\frac{3}{2}}\right) - \left(\dfrac{2}{3}\right)$

Remember that $2^{\frac{3}{2}} = \sqrt{8} = 2\sqrt{2}$

   So $I = \dfrac{2}{3}(2\sqrt{2} - 1)$

Problem-solving

You could also convert the integral back into a function of x and use the original limits.

Exercise 6C **SKILLS** PROBLEM-SOLVING

1 Use the substitutions given to find

a $\int x\sqrt{1+x}\,dx; \ u = 1 + x$

b $\int \dfrac{1 + \sin x}{\cos x}\,dx; \ u = \sin x$

c $\int \sin^3 x\,dx; \ u = \cos x$

d $\int \dfrac{2}{\sqrt{x}(x-4)}\,dx; \ u = \sqrt{x}$

e $\int \sec^2 x \tan x\sqrt{1 + \tan x}\,dx; \ u^2 = 1 + \tan x$

f $\int \sec^4 x\,dx; \ u = \tan x$

2 Use the substitutions given to find the exact values of

a $\int_0^5 x\sqrt{x+4}\,dx; \ u = x + 4$

b $\int_0^2 x(2+x)^3\,dx; \ u = 2 + x$

c $\int_0^{\frac{\pi}{2}} \sin x\sqrt{3\cos x + 1}\,dx; \ u = \cos x$

d $\int_0^{\frac{\pi}{3}} \sec x \tan x\sqrt{\sec x + 2}\,dx; \ u = \sec x$

e $\int_1^4 \dfrac{1}{\sqrt{x}(4x-1)}\,dx; \ u = \sqrt{x}$

P **3** By choosing a suitable substitution, find

a $\int x(3 + 2x)^5\,dx$

b $\int \dfrac{x}{\sqrt{1+x}}\,dx$

c $\int \dfrac{\sqrt{x^2+4}}{x}\,dx$

P **4** By choosing a suitable substitution, find the exact values of

a $\int_2^7 x\sqrt{2+x}\,dx$

b $\int_2^5 \dfrac{1}{1+\sqrt{x-1}}\,dx$

c $\int_0^{\frac{\pi}{2}} \dfrac{\sin 2\theta}{1 + \cos\theta}\,d\theta$

E **5** Using the substitution $u^2 = 4x + 1$, or otherwise, find the exact value of $\int_6^{20} \dfrac{8x}{\sqrt{4x+1}}\,dx$ **(8 marks)**

E/P **6** Use the substitution $u^2 = e^x - 2$ to show that $\int_{\ln 3}^{\ln 4} \dfrac{e^{4x}}{e^x - 2}\,dx = \dfrac{a}{b} + c\ln d$, where a, b, c and d are integers to be found. **(7 marks)**

E/P **7** Use the substitution $u = \cos x$ to show

$\int_0^{\frac{\pi}{3}} \sin^3 x \cos^2 x\,dx = \dfrac{47}{480}$ **(7 marks)**

> **Hint** Use exact trigonometric values to change the limits in x to limits in u.

E/P **8** Using a suitable trigonometric substitution for x, find $\int_{\frac{1}{2}}^{\frac{\sqrt{3}}{2}} x^2\sqrt{1-x^2}\,dx$ **(8 marks)**

Challenge

By using a substitution of the form $x = k\sin u$, show that

$\int \dfrac{1}{x^2\sqrt{9-x^2}}\,dx = -\dfrac{\sqrt{9-x^2}}{9x} + c$

6.4 Integration by parts

You can rearrange the product rule for differentiation:

$$\frac{d}{dx}(uv) = u\frac{dv}{dx} + v\frac{du}{dx}$$

$$u\frac{dv}{dx} = \frac{d}{dx}(uv) - v\frac{du}{dx}$$

$$\int u\frac{dv}{dx}\,dx = \int \frac{d}{dx}(uv)\,dx - \int v\frac{du}{dx}\,dx$$

> **Links** u and v are both functions of x.

Differentiating a function and then integrating it leaves the original function unchanged.

So, $\int \frac{d}{dx}(uv)\,dx = uv$

■ **This method is called integration by parts.** $\int u\dfrac{dv}{dx}\,dx = uv - \int v\dfrac{du}{dx}\,dx$

To use integration by parts you need to write the function you are integrating in the form $u\dfrac{dv}{dx}$

You will have to choose what to set as u and what to set as $\dfrac{dv}{dx}$

Example 9 **SKILLS** ADAPTIVE LEARNING

Find $\int x\cos x\,dx$

> **Problem-solving**
>
> For expressions like $x\cos x$, $x^2\sin x$ and $x^3 e^x$ let u equal the x^n term. When the expression involves $\ln x$, for example $x^2 \ln x$, let u equal the $\ln x$ term.

Let $I = \int x\cos x\,dx$

$u = x \quad\Rightarrow\quad \dfrac{du}{dx} = 1$

$\dfrac{dv}{dx} = \cos x \Rightarrow v = \sin x$

Using the integration by parts formula:

$I = x\sin x - \int \sin x \times 1\,dx$

$ = x\sin x + \cos x + c$

> Let $u = x$ and $\dfrac{dv}{dx} = \cos x$.
>
> Find expressions for u, v, $\dfrac{du}{dx}$ and $\dfrac{dv}{dx}$
>
> Take care to differentiate u but integrate $\dfrac{dv}{dx}$

> Notice that $\int v\dfrac{du}{dx}\,dx$ is a simpler integral than $\int u\dfrac{dv}{dx}\,dx$

Example 10 **SKILLS** PROBLEM-SOLVING

Find $\int x^2 \ln x\,dx$

Let $I = \int x^2 \ln x\,dx$

$u = \ln x \;\Rightarrow\; \dfrac{du}{dx} = \dfrac{1}{x}$

$\dfrac{dv}{dx} = x^2 \Rightarrow v = \dfrac{x^3}{3}$

> Since there is a $\ln x$ term, let $u = \ln x$ and $\dfrac{dv}{dx} = x^2$
>
> Find expressions for u, v, $\dfrac{du}{dx}$ and $\dfrac{dv}{dx}$
>
> Take care to differentiate u but integrate $\dfrac{dv}{dx}$

$$I = \frac{x^3}{3}\ln x - \int \frac{x^3}{3} \times \frac{1}{x}dx$$

$$= \frac{x^3}{3}\ln x - \int \frac{x^2}{3}dx$$

$$= \frac{x^3}{3}\ln x - \frac{x^3}{9} + c$$

— Apply the integration by parts formula.

— Simplify the $v\dfrac{du}{dx}$ term.

It is sometimes necessary to use integration by parts twice, as shown in the following example.

Example (11)   **SKILLS**   PROBLEM-SOLVING

Find $\int x^2 e^x \, dx$

There is no $\ln x$ term, so let $u = x^2$ and $\dfrac{dv}{dx} = e^x$

Find expressions for u, v, $\dfrac{du}{dx}$ and $\dfrac{dv}{dx}$

Take care to differentiate u but integrate $\dfrac{dv}{dx}$

Let $I = \int x^2 e^x \, dx$

$u = x^2 \;\Rightarrow\; \dfrac{du}{dx} = 2x$

$\dfrac{dv}{dx} = e^x \Rightarrow v = e^x$

— Apply the integration by parts formula.

So $I = x^2 e^x - \underline{\int 2x e^x dx}$

$u = 2x \;\Rightarrow\; \dfrac{du}{dx} = 2$

$\dfrac{dv}{dx} = e^x \Rightarrow v = e^x$

Notice that this integral is simpler than I but still not one you can write down. It has a similar structure to I and so you can use integration by parts again with $u = 2x$ and $\dfrac{dv}{dx} = e^x$

So $I = x^2 e^x - \left(2x e^x - \int 2 e^x dx\right)$

$= x^2 e^x - 2x e^x + \int 2 e^x dx$

$= x^2 e^x - 2x e^x + 2 e^x + c$

— Apply the integration by parts formula for a second time.

Example (12)   **SKILLS**   CREATIVITY

Evaluate $\displaystyle\int_1^2 \ln x \, dx$, leaving your answer in terms of **natural logarithms**.

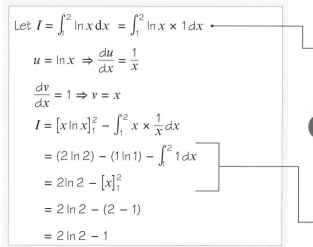

Let $I = \displaystyle\int_1^2 \ln x \, dx = \int_1^2 \ln x \times 1 \, dx$

$u = \ln x \;\Rightarrow\; \dfrac{du}{dx} = \dfrac{1}{x}$

$\dfrac{dv}{dx} = 1 \Rightarrow v = x$

$I = \left[x \ln x\right]_1^2 - \displaystyle\int_1^2 x \times \dfrac{1}{x} dx$

$= (2 \ln 2) - (1 \ln 1) - \displaystyle\int_1^2 1 \, dx$

$= 2\ln 2 - \left[x\right]_1^2$

$= 2 \ln 2 - (2 - 1)$

$= 2 \ln 2 - 1$

Write the expression to be integrated as $\ln x \times 1$ then $u = \ln x$ and $\dfrac{dv}{dx} = 1$

Remember if an expression involves $\ln x$ you should always set $u = \ln x$

Problem-solving

Apply limits to the uv term and the $\int v\dfrac{du}{dx}dx$ term separately.

Evaluate the limits on uv and remember $\ln 1 = 0$

Example (13) **SKILLS** PROBLEM-SOLVING

Using integration by parts, find $\int e^x \sin x \, dx$.

$I = \int e^x \sin x \, dx$	Label the integral I.
$u = e^x, \dfrac{du}{dx} = e^x$ $\dfrac{dv}{dx} = \sin x, v = -\cos x$	Find expressions for u, v, $\dfrac{du}{dx}$ and $\dfrac{dv}{dx}$
$I = [-e^x \cos x] - \int -e^x \cos x \, dx$	Apply the integration by parts formula.
$= [-e^x \cos x] + \int e^x \cos x \, dx$	Simplify.
$u = e^x, \dfrac{du}{dx} = e^x$ $\dfrac{dv}{dx} = \cos x, v = \sin x$	Find expressions for u, v, $\dfrac{du}{dx}$ and $\dfrac{dv}{dx}$, taking care not to switch the functions (you are still differentiating e^x).
$I = [-e^x \cos x] + \left([e^x \sin x] - \int e^x \sin x \, dx \right)$	Apply the integration by parts formula again.
$I = [-e^x \cos x + e^x \sin x] - I$	Recognise that the original integral is now on the right hand side.
$2I = [e^x (\sin x - \cos x)]$	
$I = \dfrac{e^x}{2} [\sin x - \cos x] + c$	Simplify the expression.

Exercise (6D) **SKILLS** PROBLEM-SOLVING

1 Find

a $\int x \sin x \, dx$ **b** $\int x e^x \, dx$ **c** $\int x \sec^2 x \, dx$

d $\int x \sec x \tan x \, dx$ **e** $\int \dfrac{x}{\sin^2 x} \, dx$

2 Find

a $\int 3 \ln x \, dx$ **b** $\int x \ln x \, dx$ **c** $\int \dfrac{\ln x}{x^3} \, dx$

d $\int (\ln x)^2 \, dx$ **e** $\int (x^2 + 1) \ln x \, dx$

3 Find

a $\int x^2 e^{-x} \, dx$ **b** $\int x^2 \cos x \, dx$ **c** $\int 12x^2 (3 + 2x)^5 \, dx$ **d** $\int 2x^2 \sin 2x \, dx$ **e** $\int 2x^2 \sec^2 x \tan x \, dx$

4 Evaluate

a $\int_0^{\ln 2} x e^{2x} \, dx$ **b** $\int_0^{\frac{\pi}{2}} x \sin x \, dx$ **c** $\int_0^{\frac{\pi}{2}} x \cos x \, dx$ **d** $\int_1^2 \dfrac{\ln x}{x^2} \, dx$

e $\int_0^1 4x(1 + x)^3 \, dx$ **f** $\int_0^{\pi} x \cos \dfrac{1}{4} x \, dx$ **g** $\int_0^{\frac{\pi}{3}} \sin x \ln (\sec x) \, dx$

Ⓔ **5 a** Use integration by parts to find $\int x \cos 4x \, dx$ **(3 marks)**

   b Use your answer to part **a** to find $\int x^2 \sin 4x \, dx$ **(3 marks)**

Ⓔ/ⓟ **6 a** Find $\int \sqrt{8-x} \, dx$ **(2 marks)**

   b Using integration by parts, or otherwise, show that

$$\int (x-2)\sqrt{8-x} \, dx = -\frac{2}{5}(8-x)^{\frac{3}{2}}(x+2) + c$$ **(6 marks)**

   c Hence find $\int_4^7 (x-2)\sqrt{8-x} \, dx$ **(2 marks)**

Ⓔ/ⓟ **7 a** Find $\int \sec^2 3x \, dx$ **(3 marks)**

   b Using integration by parts, or otherwise, find $\int x \sec^2 3x \, dx$ **(6 marks)**

   c Hence show that $\int_{\frac{\pi}{18}}^{\frac{\pi}{9}} x \sec^2 3x \, dx = p\pi - q \ln 3$, finding the exact values of the constants p and q.   **(4 marks)**

6.5   Partial fractions

Partial fractions can be used to integrate algebraic fractions.

Using partial fractions enables an expression that looks hard to integrate to be transformed into two or more expressions that are easier to integrate.

> **Links**   Make sure you are confident expressing algebraic fractions as partial fractions   ← **Pure 4 Section 2**

Example 14   **SKILLS** ▸ INNOVATION

Use partial fractions to find the following integrals.

a $\displaystyle\int \frac{x-5}{(x+1)(x-2)} \, dx$ **b** $\displaystyle\int \frac{8x^2 - 19x + 1}{(2x+1)(x-2)^2} \, dx$ **c** $\displaystyle\int \frac{2}{1-x^2} \, dx$

a $\dfrac{x-5}{(x+1)(x-2)} \equiv \dfrac{A}{x+1} + \dfrac{B}{x-2}$

    So $x - 5 \equiv A(x-2) + B(x+1)$

    Let $x = -1$: $-6 = A(-3)$ so $A = 2$

    Let $x = 2$: $-3 = B(3)$ so $B = -1$

    So $\displaystyle\int \dfrac{x-5}{(x+1)(x-2)} \, dx$

       $= \displaystyle\int \left(\dfrac{2}{x+1} - \dfrac{1}{x-2} \right) dx$

       $= 2 \ln|x+1| - \ln|x-2| + c$

       $= \ln \left| \dfrac{(x+1)^2}{x-2} \right| + c$

— Split the expression to be integrated into partial fractions.

— Let $x = -1$ and 2

— Rewrite the integral and integrate each term as in ← **Pure 3 Section 6.4**

— Remember to use the **modulus** when using ln in integration.

— The answer could be left in this form, but sometimes you may be asked to combine the ln terms using the rules of logarithms. ← **Pure 2 Section 3.3**

b Let $I = \int \dfrac{8x^2 - 19x + 1}{(2x + 1)(x - 2)^2} dx$

$\dfrac{8x^2 - 19x + 1}{(2x + 1)(x - 2)^2} \equiv \dfrac{A}{2x + 1} + \dfrac{B}{(x - 2)^2} + \dfrac{C}{x - 2}$

$8x^2 - 19x + 1 \equiv A(x - 2)^2 + B(2x + 1) +$
$\qquad\qquad\qquad\qquad C(2x + 1)(x - 2)$

Let $x = 2$: $-5 = 0 + 5B + 0$ so $B = -1$

Let $x = -\dfrac{1}{2}$: $12\dfrac{1}{2} = \dfrac{25}{4}A + 0 + 0$ so $A = 2$

Let $x = 0$: Then $1 = 4A + B - 2C$

So $\quad 1 = 8 - 1 - 2C$ so $C = 3$

$I = \int \left(\dfrac{2}{2x + 1} - \dfrac{1}{(x - 2)^2} + \dfrac{3}{x - 2} \right) dx$

$\quad = \dfrac{2}{2} \ln|2x + 1| + \dfrac{1}{x - 2} + 3\ln|x - 2| + c$

$\quad = \ln|2x + 1| + \dfrac{1}{x - 2} + \ln|x - 2|^3 + c$

$\quad = \ln|(2x + 1)(x - 2)^3| + \dfrac{1}{x - 2} + c$

- It is sometimes useful to label the integral as I.

- Remember the partial fraction form for a repeated factor in the denominator.

- Rewrite the intergral using the partial fractions. Note that using I saves copying the question again.

- Don't forget to divide by 2 when integrating $\dfrac{1}{2x + 1}$ and remember that the integral of $\dfrac{1}{(x - 2)^2}$ does not involve ln.

- Simplify using the laws of logarithms.

c Let $I = \int \dfrac{2}{1 - x^2} dx$

$\dfrac{2}{1 - x^2} = \dfrac{2}{(1 - x)(1 + x)} = \dfrac{A}{1 - x} + \dfrac{B}{1 + x}$

$2 = A(1 + x) + B(1 - x)$

Let $x = -1$ then $2 = 2B$ so $B = 1$

Let $x = 1$ then $2 = 2A$ so $A = 1$

So $\quad I = \int \left(\dfrac{1}{1 + x} + \dfrac{1}{1 - x} \right) dx$

$\quad = \ln|1 + x| - \ln|1 - x| + c$

$\quad = \ln\left| \dfrac{1 + x}{1 - x} \right| + c$

- Remember that $1 - x^2$ can be factorised using the difference of two squares.

- Rewrite the integral using the partial fractions.

- Notice the minus sign that comes from integrating $\dfrac{1}{1 - x}$

When the **degree of the polynomial** in the numerator is greater than or equal to the degree of the denominator, it is necessary to first divide the numerator by the denominator.

Example **15** **SKILLS** **CRITICAL THINKING**

Find $\int \dfrac{9x^2 - 3x + 2}{9x^2 - 4} dx$

Let $I = \int \dfrac{9x^2 - 3x + 2}{9x^2 - 4} dx$

$\begin{array}{r} 1 \\ 9x^2 - 4 \overline{)9x^2 - 3x + 2} \\ 9x^2 \qquad -4 \\ \hline -3x + 6 \end{array}$

- First divide the numerator by $9x^2 - 4$

- $9x^2 \div 9x^2$ gives 1, so put this on top and subtract $1 \times (9x^2 - 4)$. This leaves a remainder of $-3x + 6$

so $I = \int \left(1 + \dfrac{6 - 3x}{9x^2 - 4}\right) dx$

$\dfrac{6 - 3x}{9x^2 - 4} \equiv \dfrac{A}{3x - 2} + \dfrac{B}{3x + 2}$ ———— Factorise $9x^2 - 4$ and then split into partial fractions.

Let $x = -\dfrac{2}{3}$ then $8 = -4B$ so $B = -2$

Let $x = \dfrac{2}{3}$ then $4 = 4A$ so $A = 1$

So $I = \int \left(1 + \dfrac{1}{3x - 2} - \dfrac{2}{3x + 2}\right) dx$ ———— Rewrite the integral using the partial fractions.

$= x + \dfrac{1}{3} \ln|3x - 2| - \dfrac{2}{3} \ln|3x + 2| + c$ ———— Integrate and don't forget the $\dfrac{1}{3}$

$= x + \dfrac{1}{3} \ln\left|\dfrac{3x - 2}{(3x + 2)^2}\right| + c$ ———— Simplify using the laws of logarithms.

Exercise **6E**   **SKILLS**   INNOVATION

1 Use partial fractions to integrate

   a $\dfrac{3x + 5}{(x + 1)(x + 2)}$
   b $\dfrac{3x - 1}{(2x + 1)(x - 2)}$
   c $\dfrac{2x - 6}{(x + 3)(x - 1)}$
   d $\dfrac{3}{(2 + x)(1 - x)}$

2 Find

   a $\displaystyle\int \dfrac{2(x^2 + 3x - 1)}{(x + 1)(2x - 1)} dx$
   b $\displaystyle\int \dfrac{x^3 + 2x^2 + 2}{x(x + 1)} dx$
   c $\displaystyle\int \dfrac{x^2}{x^2 - 4} dx$
   d $\displaystyle\int \dfrac{x^2 + x + 2}{3 - 2x - x^2} dx$

E/P **3** $f(x) = \dfrac{4}{(2x + 1)(1 - 2x)}$, $x \neq \pm\dfrac{1}{2}$

   a Given that $f(x) = \dfrac{A}{2x + 1} + \dfrac{B}{1 - 2x}$, find the value of the constants A and B.   **(3 marks)**

   b Hence find $\int f(x)\,dx$, writing your answer as a single logarithm.   **(4 marks)**

   c Find $\displaystyle\int_1^2 f(x)\,dx$, giving your answer in the form $\ln k$ where k is a rational constant.   **(2 marks)**

E/P **4** $f(x) = \dfrac{17 - 5x}{(3 + 2x)(2 - x)^2}$, $-\dfrac{3}{2} < x < 2$

   a Express $f(x)$ in partial fractions.   **(4 marks)**

   b Hence find the exact value of $\displaystyle\int_0^1 \dfrac{17 - 5x}{(3 + 2x)(2 - x)^2} dx$, writing your answer in the form $a + \ln b$, where a and b are constants to be found.   **(5 marks)**

E/P **5** $f(x) = \dfrac{9x^2 + 4}{9x^2 - 4}$, $x \neq \pm\dfrac{2}{3}$

   a Given that $f(x) = A + \dfrac{B}{3x - 2} + \dfrac{C}{3x + 2}$, find the values of the constants A, B and C.   **(4 marks)**

   b Hence find the exact value of

   $\displaystyle\int_{-\frac{1}{3}}^{\frac{1}{3}} \dfrac{9x^2 + 4}{9x^2 - 4} dx$, writing your answer in the

   form $a + b \ln c$, where a, b and c are
   rational numbers to be found.   **(5 marks)**

Problem-solving

Simplify the integral as much as possible before substituting your p limits.

(E/P) **6** $f(x) = \dfrac{6 + 3x - x^2}{x^3 + 2x^2}$, $x > 0$

    a Express $f(x)$ in partial fractions.   **(4 marks)**

    b Hence find the exact value of $\displaystyle\int_2^4 \dfrac{6 + 3x - x^2}{x^3 + 2x^2}\,dx$, writing your answer in the form $a + \ln b$,

    where a and b are rational numbers to be found.   **(5 marks)**

(E/P) **7** $\dfrac{32x^2 + 4}{(4x + 1)(4x - 1)} \equiv A + \dfrac{B}{4x + 1} + \dfrac{C}{4x - 1}$

    a Find the value of the constants A, B and C.   **(4 marks)**

    b Hence find the exact value of $\displaystyle\int_1^2 \dfrac{32x^2 + 4}{(4x + 1)(4x - 1)}\,dx$ writing your answer in the form

    $2 + k \ln m$, giving the values of the rational constants k and m.   **(5 marks)**

6.6   Solving differential equations

Integration can be used to solve differential equations. In this chapter you will solve first order differential equations by **separating the variables**.

- When $\dfrac{dy}{dx} = f(x)g(y)$ you can write

$$\int \frac{1}{g(y)}\,dy = \int f(x)\,dx$$

> **Notation**   A first order differential equation contains nothing higher than a first order derivative, for example $\dfrac{dy}{dx}$
>
> A second order differential equation would have a term that contains a second order derivative, for example $\dfrac{d^2y}{dx^2}$

The solution to a differential equation will be a function.
When you integrate to solve a differential equation you still need to include a constant of integration. This gives the **general solution** to the differential equation. It represents a **family** of solutions, all with different constants. Each of these solutions satisfies the original differential equation.

For the first order differential equation $\dfrac{dy}{dx} = 12x^2 - 1$, the general solution is $y = 4x^3 - x + c$, or $y = x(2x - 1)(2x + 1) + c$

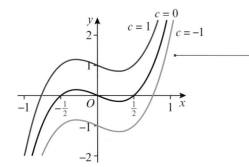

Each of these curves represents a **particular solution** of the differential equation, for different values of the constant c. Together, the curves form a **family of solutions**.

> **Online**   Explore families of solutions using technology.

Example **16**   **SKILLS**   ADAPTIVE LEARNING

Find a general solution to the differential equation $(1 + x^2)\dfrac{dy}{dx} = x \tan y$

$$\frac{dy}{dx} = \frac{x}{1 + x^2}\tan y$$

Write the equation in the form $\dfrac{dy}{dx} = f(x)g(y)$

$$\int \frac{1}{\tan y}\,dy = \int \frac{x}{1 + x^2}\,dx$$

$$\int \cot y\,dy = \int \frac{x}{1 + x^2}\,dx$$

Now **separate the variables**:

$$\frac{1}{g(y)}\,dy = f(x)\,dx$$

$$\ln|\sin y| = \frac{1}{2}\ln|1 + x^2| + c$$

or $\quad \ln|\sin y| = \dfrac{1}{2}\ln|1 + x^2| + \ln k$

$$\ln|\sin y| = \ln|k\sqrt{1 + x^2}|$$

so $\quad \sin y = k\sqrt{1 + x^2}$

Use $\cot y = \dfrac{1}{\tan y}$

$$\int \cot x\,dx = \ln|\sin x| + c$$

Don't forget the $+c$ which can be written as $\ln k$.

Finally remove the ln. Sometimes you might be asked to give your answer in the form $y = f(x)$ This question did not specify that so it is acceptable to give the answer in this form.

Combining logs.

Sometimes you are interested in one specific solution to a differential equation. You can find a **particular solution** to a first-order differential equation if you know **one point** on the curve. This is sometimes called a **boundary condition**.

Example **17**   **SKILLS**   CRITICAL THINKING

Find the particular solution to the differential equation

$$\frac{dy}{dx} = \frac{-3(y - 2)}{(2x + 1)(x + 2)}$$

given that $x = 1$ when $y = 4$. Leave your answer in the form $y = f(x)$

Hint The boundary condition in this question is that $x = 1$ when $y = 4$

$$\int \frac{1}{y - 2}\,dy = \int \frac{-3}{(2x + 1)(x + 2)}\,dx$$

$$\frac{-3}{(2x + 1)(x + 2)} \equiv \frac{A}{(2x + 1)} + \frac{B}{(x + 2)}$$

First separate the variables. Make sure the function on the left-hand side is in terms of y only, and the function on the right-hand side is in terms of x only.

$$-3 = A(x + 2) + B(2x + 1)$$

Let $x = -2$: $\quad -3 = -3B \;$ so $\; B = 1$

Let $x = -\dfrac{1}{2}$: $\quad -3 = \dfrac{3}{2}A \;$ so $\; A = -2$

Convert the fraction on the RHS to partial fractions.

So

$$\int \frac{1}{y - 2}\,dy = \int \left(\frac{1}{x + 2} - \frac{2}{2x + 1}\right)dx$$

Rewrite the integral using the partial fractions.

$$\ln|y - 2| = \ln|x + 2| - \ln|2x + 1| + \ln k$$ ——— Integrate and use $+\ln k$ instead of $+c$

$$\ln|y - 2| = \ln\left|\frac{k(x + 2)}{2x + 1}\right|$$ ——— Combine ln terms.

$$y - 2 = k\left(\frac{x + 2}{2x + 1}\right)$$ ——— Remove ln.

$$4 - 2 = k\left(\frac{1 + 2}{2 + 1}\right) \Rightarrow k = 2$$ ——— Use the condition $x = 1$ when $y = 4$ by substituting these values into the general solution and solving to find k.

So $$y = 2 + 2\left(\frac{x + 2}{2x + 1}\right)$$

$$y = 3 + \frac{3}{2x + 1}$$ ——— Substitute $k = 2$ and write the answer in the form $y = f(x)$ as requested.

Exercise **6F**   **SKILLS**   CRITICAL THINKING

1 Find general solutions to the following differential equations.
 Give your answers in the form $y = f(x)$

 a $\dfrac{dy}{dx} = (1 + y)(1 - 2x)$ **b** $\dfrac{dy}{dx} = y \tan x$

 c $\cos^2 x \dfrac{dy}{dx} = y^2 \sin^2 x$ **d** $\dfrac{dy}{dx} = 2e^{x - y}$

2 Find particular solutions to the following differential equations using the given
 boundary conditions.

 a $\dfrac{dy}{dx} = \sin x \cos^2 x; \ y = 0, \ x = \dfrac{\pi}{3}$ **b** $\dfrac{dy}{dx} = \sec^2 x \sec^2 y; \ y = 0, \ x = \dfrac{\pi}{4}$

 c $\dfrac{dy}{dx} = 2\cos^2 y \cos^2 x; \ y = \dfrac{\pi}{4}, \ x = 0$ **d** $\sin y \cos x \dfrac{dy}{dx} = \dfrac{\cos y}{\cos x}, \ y = 0, \ x = 0$

3 **a** Find the general solution to the differential equation
 $x^2 \dfrac{dy}{dx} = y + xy$, giving your answer in the form $y = g(x)$

 b Find the particular solution to the differential equation that
 satisfies the boundary condition $y = e^4$ at $x = -1$

 > **Hint** Begin by factorising the right-hand side of the equation.

(E) 4 Given that $x = 0$ when $y = 0$, find the particular solution to the differential equation
 $(2y + 2yx)\dfrac{dy}{dx} = 1 - y^2$, giving your answer in the form $y = g(x)$ **(6 marks)**

(E/P) 5 Find the general solution to the differential equation $e^{x+y}\dfrac{dy}{dx} = 2x + xe^y$, giving your answer
 in the form $\ln|g(y)| = f(x)$ **(6 marks)**

(E) 6 Find the particular solution to the differential equation $(1 - x^2)\dfrac{dy}{dx} = xy + y$, with boundary
 condition $y = 6$ at $x = 0.5$ Give your answer in the form $y = f(x)$ **(8 marks)**

(E) **7** Find the particular solution to the differential equation $(1 + x^2)\dfrac{dy}{dx} = x - xy^2$, with boundary condition $y = 2$ at $x = 0$ Give your answer in the form $y = f(x)$ **(8 marks)**

(E) **8** Find the particular solution to the differential equation $\dfrac{dy}{dx} = xe^{-y}$, with boundary condition $y = \ln 2$ at $x = 4$. Give your answer in the form $y = f(x)$ **(8 marks)**

(E/P) **9** Find the particular solution to the differential equation $\dfrac{dy}{dx} = \cos^2 y + \cos 2x \cos^2 y$, with boundary condition $y = \dfrac{\pi}{4}$ at $x = \dfrac{\pi}{4}$ Give your answer in the form $\tan y = f(x)$ **(8 marks)**

(E) **10** Given that $y = 1$ at $x = \dfrac{\pi}{2}$, solve the differential equation $\dfrac{dy}{dx} = xy \sin x$ **(6 marks)**

(E) **11 a** Find $\displaystyle\int \dfrac{3x + 4}{x}\,dx,\ x > 0$ **(2 marks)**

   b Given that $y = 16$ at $x = 1$, solve the differential equation $\dfrac{dy}{dx} = \dfrac{3x\sqrt{y} + 4\sqrt{y}}{x}$
   giving your answer in the form $y = f(x)$ **(6 marks)**

(E) **12 a** Express $\dfrac{8x - 18}{(3x - 8)(x - 2)}$ in partial fractions.   **(3 marks)**

   b Given that $x \geqslant 3$, find the general solution to the differential equation
   $(x - 2)(3x - 8)\dfrac{dy}{dx} = (8x - 18)y$ **(5 marks)**

   c Hence find the particular solution to this differential equation that satisfies
   $y = 8$ at $x = 3$ giving your answer in the form $y = f(x)$.   **(4 marks)**

(P) **13 a** Find the general solution of $\dfrac{dy}{dx} = 2x - 4$

   b On the same axes, sketch three different particular solutions to this differential equation.

(E/P) **14 a** Find the general solution to the differential equation $\dfrac{dy}{dx} = -\dfrac{1}{(x + 2)^2}$ **(3 marks)**

   b On the same axes, sketch three different particular solutions to this differential equation.   **(3 marks)**

   c Write down the particular solution that passes through the point $(8, 3.1)$ **(1 mark)**

(E/P) **15 a** Show that the general solution to the differential equation $\dfrac{dy}{dx} = -\dfrac{x}{y}$ can be written in the form $x^2 + y^2 = c$ **(3 marks)**

   b On the same axes, sketch three different particular solutions to this differential equation.   **(3 marks)**

   c Write down the particular solution that passes through the point $(0, 7)$ **(1 mark)**

6.7 Modelling with differential equations

Differential equations can be used to model real-life situations.

Example 18 SKILLS INNOVATION

The rate of increase of a population P of microorganisms at time t, in hours, is given by

$$\frac{dP}{dt} = 3P, k > 0$$

Initially the population was of size 8.

a Find a model for P in the form $P = Ae^{3t}$ stating the value of A.

b Find, to the nearest hundred, the size of the population at time $t = 2$

c Find the time at which the population will be 1000 times its starting value.

d State one limitation of this model for large values of t.

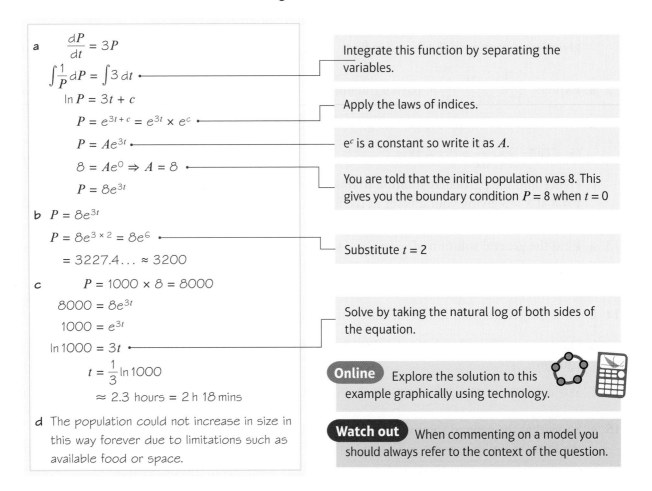

a $\dfrac{dP}{dt} = 3P$

$\displaystyle\int \dfrac{1}{P}\,dP = \int 3\,dt$ —— Integrate this function by separating the variables.

$\ln P = 3t + c$ —— Apply the laws of indices.

$P = e^{3t+c} = e^{3t} \times e^{c}$ —— e^c is a constant so write it as A.

$P = Ae^{3t}$

$8 = Ae^{0} \Rightarrow A = 8$ —— You are told that the initial population was 8. This gives you the boundary condition $P = 8$ when $t = 0$

$P = 8e^{3t}$

b $P = 8e^{3t}$

$P = 8e^{3 \times 2} = 8e^{6}$ —— Substitute $t = 2$

$= 3227.4\ldots \approx 3200$

c $P = 1000 \times 8 = 8000$

$8000 = 8e^{3t}$

$1000 = e^{3t}$ —— Solve by taking the natural log of both sides of the equation.

$\ln 1000 = 3t$

$t = \dfrac{1}{3}\ln 1000$

$\approx 2.3 \text{ hours} = 2\text{ h } 18 \text{ mins}$

Online Explore the solution to this example graphically using technology.

d The population could not increase in size in this way forever due to limitations such as available food or space.

Watch out When commenting on a model you should always refer to the context of the question.

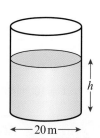

Example 19   **SKILLS**   INTERPRETATION

Water in a manufacturing plant is held in a large cylindrical tank of diameter 20 m.
Water flows out of the bottom of the tank through a tap at a rate proportional
to the cube root of the volume.

a Show that t minutes after the tap is opened, $\dfrac{dh}{dt} = -k\sqrt[3]{h}$ for some constant k.

b Show that the general solution to this differential equation may be written
as $h = (P - Qt)^{\frac{3}{2}}$, where P and Q are constants.

Initially the height of the water is 27 m. 10 minutes later, the height is 8 m.

c Find the values of the constants P and Q.

d Find the time in minutes when the water is at a depth of 1 m.

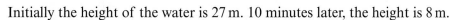

a $V = \pi r^2 h = 100\pi h$

$\dfrac{dV}{dh} = 100\pi$

$\dfrac{dV}{dt} = -c\sqrt[3]{V}$

$\quad = -c\sqrt[3]{100\pi h}$

$\dfrac{dh}{dt} = \dfrac{dh}{dV} \times \dfrac{dV}{dt}$

$\dfrac{dh}{dt} = \dfrac{1}{100\pi} \times (-c\sqrt[3]{100\pi h})$

$\quad = \left(\dfrac{-c\sqrt[3]{100\pi}}{100\pi}\right)\sqrt[3]{h}$

So $\dfrac{dh}{dt} = -k\sqrt[3]{h}$, where $k = \dfrac{c\sqrt[3]{100\pi}}{100\pi}$

b $\displaystyle\int h^{-\frac{1}{3}}\, dh = -\int k\, dt$

$\dfrac{3}{2}h^{\frac{2}{3}} = -kt + c$

$h^{\frac{2}{3}} = -\dfrac{2}{3}kt + \dfrac{2}{3}c$

$h^{\frac{2}{3}} = -Qt + P$

$h = (P - Qt)^{\frac{3}{2}}$

c $t = 0, h = 27$

$27 = P^{\frac{3}{2}} \Rightarrow P = 9$

$t = 10, h = 8$

$8 = (9 - 10Q)^{\frac{3}{2}}$

$4 = 9 - 10Q$

$Q = \dfrac{1}{2}$

Use the formula for the volume of a cylinder. The diameter is 20, so the radius is 10.

Problem-solving

You need to use the information given in the question to construct a mathematical model. Water flows out at a rate proportional to the cube root of the volume.

$\dfrac{dV}{dt}$ is negative as the water is flowing out of the tank, so the volume is decreasing.

Use the chain rule to find $\dfrac{dh}{dt}$

Substitute for $\dfrac{dh}{dV}$ and $\dfrac{dV}{dt}$

$\dfrac{dh}{dV} = \dfrac{1}{\frac{dV}{dh}} = \dfrac{1}{100\pi}$

c was the constant of proportionality and π is constant so $\dfrac{c \times \sqrt[3]{100\pi}}{100\pi} = k$ is a constant.

Integrate this function by separating the variables.

Let $Q = \dfrac{2}{3}k$ and $P = \dfrac{2}{3}c$

Use the boundary conditions to find the values of P and Q. If there are two boundary conditions then you should consider the initial condition (when $t = 0$) first.

d $h = \left(9 - \frac{1}{2}t\right)^{\frac{3}{2}}$

$1 = \left(9 - \frac{1}{2}t\right)^{\frac{3}{2}}$

$1 = 9 - \frac{1}{2}t$

$t = 16$ minutes

Set $h = 1$ and solve the resulting equation to find the corresponding value of t.

Exercise 6G   SKILLS   ANALYSIS

(E/P) 1 The rate of increase of a population P of rabbits at time t, in years, is given by $\frac{dP}{dt} = kP$, $k > 0$
Initially the population was of size 200.
  a Solve the differential equations giving P in terms of k and t. **(3 marks)**
  b Given that $k = 3$, find the time taken for the population to reach 4000. **(4 marks)**
  c State a limitation of this model for large values of t. **(1 mark)**

(E/P) 2 The mass M at time t of the leaves of a certain plant varies according to the differential equation
$$\frac{dM}{dt} = M - M^2$$
  a Given that at time $t = 0$, $M = 0.5$, find an expression for M in terms of t. **(5 marks)**
  b Find a value of M when $t = \ln 2$ **(2 marks)**
  c Explain what happens to the value of M as t increases. **(1 mark)**

(E/P) 3 The thickness of ice x, in cm, on a pond is increasing at a rate that is inversely proportional to the square of the existing thickness of ice. Initially, the thickness is 1 cm. After 20 days, the thickness is 2 cm.
  a Show that the thickness of ice can be modelled by the equation $x = \sqrt[3]{\frac{7}{20}t + 1}$ **(7 marks)**
  b Find the time taken for the ice to increase in thickness from 2 cm to 3 cm. **(2 marks)**

(E/P) 4 A mug of tea, with a temperature $T°C$ is made and left to cool in a room with a temperature of 25 °C. The rate at which the tea cools is proportional to the difference in temperature between the tea and the room.
  a Show that this process can be described by the differential equation $\frac{dT}{dt} = -k(T - 25)$ explaining why k is a positive constant. **(3 marks)**
  Initially the tea is at a temperature of 85 °C. 10 minutes later the tea is at 55 °C.
  b Find the temperature, to 1 decimal place, of the tea after 15 minutes. **(7 marks)**

(E/P) 5 The rate of change of the surface area of a drop of oil, A mm², at time t minutes can be
  modelled by the equation $\frac{dA}{dt} = \frac{A^{\frac{3}{2}}}{10t^2}$
  Given that the surface area of the drop is 1 mm² at $t = 1$
  a find an expression for A in terms of t **(7 marks)**
  b show that the surface area of the drop cannot exceed $\frac{400}{361}$ mm². **(2 marks)**

(E/P) **6** A bath tub is modelled as a **cuboid** with a **base area** of 6000 cm². Water flows into the bath tub from a tap at a rate of 12 000 cm³/min. At time t minutes, the depth of water in the bath tub is h cm. Water leaves the bottom of the bath through an open plughole at a rate of $500h$ cm³/min.

  a Show that t minutes after the tap has been opened, $60\dfrac{dh}{dt} = 120 - 5h$ **(3 marks)**

  When $t = 0$, $h = 6$ cm

  b Find the value of t when $h = 10$ cm   **(5 marks)**

(E/P) **7** **a** Express $\dfrac{1}{P(10\,000 - P)}$ using partial fractions.   **(3 marks)**

  The deer population, P, in a reservation can be modelled by the differential equation

$$\frac{dP}{dt} = \frac{1}{200}P(10\,000 - P)$$

  where t is the time in years since the study began.

  b Given that the initial deer population is 2500, solve the differential equation giving your answer in the form $P = \dfrac{a}{b + ce^{-50t}}$ **(6 marks)**

  c Find the maximum deer population according to the model.   **(2 marks)**

(E/P) **8** Liquid is pouring into a container at a constant rate of 40 cm³ s⁻¹ and is leaking from the container at a rate of $\frac{1}{4}V$ cm³ s⁻¹, where V cm³ is the volume of liquid in the container.

  a Show that $-4\dfrac{dV}{dt} = V - 160$ **(2 marks)**

  Given that $V = 5000$ when $t = 0$

  b find the solution to the differential equation in the form $V = a + be^{-\frac{1}{4}t}$, where a and b are constants to be found   **(7 marks)**

  c write down the limiting value of V as $t \to \infty$ **(1 mark)**

(E/P) **9** Fossils are aged using a process called carbon dating. The amount of carbon remaining in a fossil, R, decreases over time, t, measured in years. The rate of decrease of carbon is proportional to the remaining carbon.

  a Given that initially the amount of carbon is R_0, show that $R = R_0e^{-kt}$ **(4 marks)**

  It is known that the half-life of carbon is 5730 years. This means that after 5730 years the amount of carbon remaining has reduced by half.

  b Find the exact value of k.   **(3 marks)**

  c A fossil is found with 10% of its expected carbon remaining. Determine the age of the fossil to the nearest year.   **(3 marks)**

Chapter review **6** **SKILLS** **DECISION MAKING**

(E) **1** A curve C is represented by the parametric equations

$x = 1 - t^2$, $y = t^3 + 1$, $-2 \leq t \leq 0$

Determine the area under the curve for the given interval. **(6 marks)**

(E) **2** A parametric curve is represented by $x = \ln(t + 2)$, $y = 4t$. Find the area under the curve from $t = 3$ to $t = 13$, giving your answer in the form $a + b \ln c$, where a, b, c are integers. **(7 marks)**

(E) **3** The curve shown in the diagram has equation $y = x^2\sqrt{9 - x^2}$
The finite region R is bounded by the curve and the x-axis.
The region is rotated through 2π radians about the x-axis
to generate a solid of revolution. Find the exact value of the
volume of the solid that is generated. **(5 marks)**

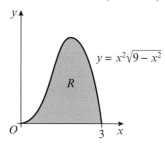

(P) **4** The diagram shows the curve with equation $2y^2 - 6\sqrt{x} + 3 = 0$
The shaded region is bounded by the curve and the line $x = 4$
a Find the value of x at the point where the curve cuts
the x-axis.
The region is rotated about the x-axis to generate a solid of revolution.
b Find the volume of the solid generated.

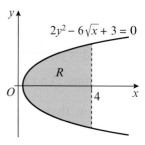

(E) **5** $f(x) = x^2 + 4x + 4$, $x \geqslant -2$
The diagram shows the finite region R bounded by the curve
$y = f(x)$, the y-axis and the lines $y = 4$ and $y = 9$
a Show that the equation $y = f(x)$ can be written as
$x^2 = 4 - 4\sqrt{y} + y$ **(2 marks)**
b The region R is rotated through 2π radians about the y-axis.
Find the exact volume of the solid generated. **(5 marks)**

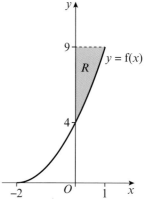

(E) **6** The diagram shows the shaded region bounded by the curve with
equation $y = x^2 + 3$, the line $x = 1$, the x-axis and the y-axis.

Find the volume generated when the region is rotated through
2π radians about the x-axis **(5 marks)**

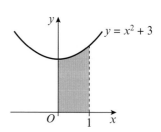

E/P **7** The diagram shows the curve with equation $y = \frac{1}{4}x(x + 1)^2$ and the
line with equation $3x + 4y = 24$. The line and the curve intersect
at the point A.

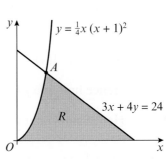

$y = \frac{1}{4}x(x + 1)^2$

A

R

$3x + 4y = 24$

a Show that the coordinates of the point A are $(2, 4.5)$ **(2 marks)**

The shaded region R is bounded by the curve, the line and the
x-axis. The region is rotated through 2π radians about the x-axis.

b Find the exact volume of solid generated. **(6 marks)**

8 The curve represented by the parametric equations $x = t^{\frac{1}{2}}$, $y = 2t^{\frac{1}{2}}$ is rotated about the x-axis
between the values $t = 1$ and $t = 4$

Determine the volume generated, giving your answer in an exact form.

9 A curve C is represented by the parametric equations $x = t^3$, $y = 3t$. The curve is then rotated
about the x-axis to form a solid. Given that the curve is rotated between the values $t = 0.4$ and
$t = 0.5$ find the volume generated, to 3 significant figures.

10 Given that the curve C, represented by the parametric equations $x = 3t^3$, $y = \frac{1}{3}t^{-\frac{3}{2}}$, is rotated
about the x-axis between the values of $t = 8$ to $t = 128$, find the exact volume generated, giving
your answer in the form $\pi \ln b$, where b is an integer.

11 By choosing a suitable method of integration, find

a $\int x\sqrt{4x - 1}\, dx$ **b** $\int x \ln x\, dx$ **c** $\int \frac{4 \sin x \cos x}{4 - 8 \sin^2 x}\, dx$

12 By choosing a suitable method, evaluate the following **definite integrals**.
Write your answers as exact values.

a $\int_0^{\frac{\pi}{4}} x \sec^2 x\, dx$ **b** $\int_1^4 \frac{4}{16x^2 + 8x - 3}\, dx$ **c** $\int_0^{\ln 2} \frac{1}{1 + e^x}\, dx$

E/P **13 a** Show that $\int_1^e \frac{1}{x^2} \ln x\, dx = 1 - \frac{2}{e}$ **(5 marks)**

b Given that $p > 1$, show that $\int_1^p \frac{1}{(x + 1)(2x - 1)}\, dx = \frac{1}{3} \ln \frac{4p - 2}{p + 1}$ **(5 marks)**

E **14 a** Using the substitution $t^2 = x + 1$, where $x > -1$ find $\int \frac{x}{\sqrt{x + 1}}\, dx$ **(5 marks)**

b Use your answer to part a to evaluate $\int_0^3 \frac{x}{\sqrt{x + 1}}\, dx$ **(2 marks)**

E **15 a** Use integration by parts to find $\int x \sin 8x\, dx$ **(4 marks)**

b Use your answer to part **a** to find $\int x^2 \cos 8x\, dx$ **(4 marks)**

(E/P) **16** $f(x) = \dfrac{5x^2 - 8x + 1}{2x(x-1)^2}$

   a Given that $f(x) = \dfrac{A}{x} + \dfrac{B}{x-1} + \dfrac{C}{(x-1)^2}$ find the values of the constants A, B and C. **(4 marks)**

   b Hence find $\int f(x)\,dx$ **(4 marks)**

   c Hence show that $\displaystyle\int_4^9 f(x)\,dx = \ln\left(\dfrac{32}{3}\right) - \dfrac{5}{24}$ **(4 marks)**

(E/P) **17** **a** Find $\int x^2 \ln 2x\,dx$ **(6 marks)**

   b Hence show that the exact value of $\displaystyle\int_{\frac{1}{2}}^3 x^2 \ln 2x\,dx$ is $9\ln 6 - \dfrac{215}{72}$ **(4 marks)**

(E) **18** **a** Find $\int xe^{-x}\,dx$ **(4 marks)**

   b Given that $y = \dfrac{\pi}{4}$ at $x = 0$, solve the differential equation

$$e^x \frac{dy}{dx} = \frac{x}{\sin 2y}$$ **(4 marks)**

(E) **19** **a** Find $\int x \sin 2x\,dx$ **(5 marks)**

   b Given that $y = 0$ at $x = \dfrac{\pi}{4}$, solve the differential equation $\dfrac{dy}{dx} = x \sin 2x \cos^2 y$ **(5 marks)**

(E/P) **20** **a** Obtain the general solution to the differential equation

$$\frac{dy}{dx} = xy^2, \; y > 0$$ **(3 marks)**

   b Given also that $y = 1$ at $x = 1$, show that

$$y = \frac{2}{3 - x^2}, \; -\sqrt{3} < x < \sqrt{3}$$

     is a particular solution to the differential equation. **(3 marks)**

     The curve C has equation $y = \dfrac{2}{3 - x^2}$, $x \neq \pm\sqrt{3}$

   c Write down the gradient of C at the point $(1, 1)$ **(1 mark)**

   d Hence write down an equation of the tangent to C at the points $(1, 1)$, and find the coordinates of the point where it again meets the curve. **(4 marks)**

(E) **21** **a** Using the substitution $u = 1 + 2x$, or otherwise, find

$$\int \frac{4x}{(1 + 2x)^2}\,dx, \; x \neq -\frac{1}{2}$$ **(5 marks)**

   b Given that $y = \dfrac{\pi}{4}$ when $x = 0$, solve the differential equation

$$(1 + 2x)^2 \frac{dy}{dx} = \frac{x}{\sin^2 y}$$ **(5 marks)**

(E/P) **22** The diagram shows the curve with equation $y = xe^{2x}$, $-\frac{1}{2} \leq x \leq \frac{1}{2}$

The finite region R_1 bounded by the curve, the x-axis and the

line $x = -\frac{1}{2}$ has area A_1.

The finite region R_2 bounded by the curve, the x-axis and the

line $x = \frac{1}{2}$ has area A_2.

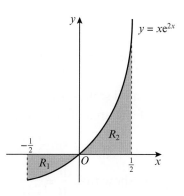

a Find the exact values of A_1 and A_2 by integration.   **(6 marks)**

b Show that $A_1 : A_2 = (e - 2) : e$ **(4 marks)**

(E) **23 a** Find $\int x^2 e^{-x} \, dx$ **(5 marks)**

b Use your answer to part **a** to find the solution to the differential equation $\dfrac{dy}{dx} = x^2 e^{3y - x}$

given that $y = 0$ when $x = 0$. Express your answer in the form $y = f(x)$ **(7 marks)**

(E) **24 a** Given that

$$\frac{x^2}{x^2 - 1} \equiv A + \frac{B}{x - 1} + \frac{C}{x + 1}$$

find the values of the constants A, B and C.   **(4 marks)**

b Given that $x = 2$ at $t = 1$, solve the differential equation

$$\frac{dx}{dt} = 2 - \frac{2}{x^2}, \; x > 1$$

You do not need to simplify your final answer.   **(7 marks)**

(E/P) **25** The rate, in $cm^3 \, s^{-1}$, at which oil is leaking from an engine sump at any time t seconds is
proportional to the volume of oil, $V \, cm^3$, in the sump at that instant. At time $t = 0$, $V = A$

a By forming and integrating a differential equation, show that $V = Ae^{-kt}$
where k is a positive constant.   **(5 marks)**

b Sketch a graph to show the relation between V and t.   **(2 marks)**

c Given further that $V = \frac{1}{2} A$ at $t = T$, show that $kT = \ln 2$ **(3 marks)**

(E/P) **26 a** Show that the general solution to the differential equation $\dfrac{dy}{dx} = \dfrac{x}{k - y}$ can be

written in the form $x^2 + (y - k)^2 = c$ **(4 marks)**

b Describe the family of curves that satisfy this differential equation when $k = 2$ **(2 marks)**

(E) **27 a** Find $\int x (1 + 2x^2)^5 \, dx$ **(3 marks)**

b Given that $y = \dfrac{\pi}{8}$ at $x = 0$, solve the differential equation $\dfrac{dy}{dx} = x(1 + 2x^2)^5 \cos^2 2y$ **(5 marks)**

(E/P) **28** By using an appropriate trigonometric substitution, find $\int \dfrac{1}{1 + x^2} dx$ **(5 marks)**

(E/P) **29** Obtain the solution to $x(x + 2)\dfrac{dy}{dx} = y$, $y > 0$, $x > 0$

for which $y = 2$ at $x = 2$, giving your answer in the form $y^2 = f(x)$ **(7 marks)**

(E/P) **30** An oil spill is modelled as a circular disc with radius r km and area A km². The rate of increase of the area of the oil spill, in km²/day at time t days after it occurs is modelled as:

$$\frac{dA}{dt} = k \sin\left(\frac{t}{3\pi}\right), \ 0 \le t \le 12$$

a Show that $\dfrac{dr}{dt} = \dfrac{k}{2\pi r} \sin\left(\dfrac{t}{3\pi}\right)$ **(2 marks)**

Given that the radius of the spill at time $t = 0$ is 1 km, and the radius of the spill at time $t = \pi^2$ is 2 km

b find an expression for r^2 in terms of t **(7 marks)**

c find the time, in days and hours to the nearest hour, after which the radius of the spill is 1.5 km. **(3 marks)**

Challenge

Given $f(x) = x^2 - x - 2$, find

a $\displaystyle\int_{-3}^{3} |f(x)| \, dx$ **b** $\displaystyle\int_{-3}^{3} f(|x|) \, dx$

Hint Draw a sketch of each function.

Summary of key points

1 The area under a curve represented by the parametric equations $x = f(t)$, $y = g(t)$ is given by: area $= \displaystyle\int_{t_1}^{t_2} g(t)f'(t)dt$

2 The volume of revolution formed when $y = f(x)$ is rotated about the x-axis between $x = a$ and $x = b$ is given by: volume $= \pi\displaystyle\int_a^b y^2 \, dx$

3 Sometimes you can simplify an integral by changing the variable. This process is similar to using the chain rule in differentiation and is called **integration by substitution**.

4 The **integration by parts** formula is given by: $\displaystyle\int u \frac{dv}{dx} \, dx = uv - \int v \frac{du}{dx} \, dx$

5 Partial fractions can be used to integrate algebraic fractions.

6 When $\dfrac{dy}{dx} = f(x)g(y)$ you can write: $\displaystyle\int \frac{1}{g(y)} \, dy = \int f(x) \, dx$

7.1
7.2
7.3
7.4
7.5
7.6
7.7

7 VECTORS

Learning objectives

After completing this chapter you should be able to:

* Use vectors in two dimensions → **pages 98–102**
* Use column vectors and carry out arithmetic operations on vectors → **pages 102–105**
* Calculate the magnitude and direction of a vector → **pages 106–108**
* Use vectors in three dimensions → **pages 109–113**
* Use vectors to solve geometric problems → **pages 114–121**
* Understand and use position vectors → **pages 121–123**
* Understand 3D Cartesian coordinates → **pages 123–125**
* Understand and use the vector form of the equation of a straight line in three dimensions → **pages 125–131**
* Determine whether two lines meet and determine the point of intersection → **pages 131–134**
* Calculate the scalar product for two 3D vectors → **pages 134–140**

Prior knowledge check

1 Write the column vector for the translation of shape:

   a A to B

   b A to C

   c A to D

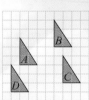

   ← **International GCSE Mathematics**

2 P divides the line AB in the ratio $AP : PB = 7 : 2$

   Find

   a $\dfrac{AP}{AB}$ **b** $\dfrac{PB}{AB}$ **c** $\dfrac{AP}{PB}$

   ← **International GCSE Mathematics**

3 Find x to one decimal place:

   a 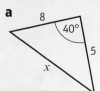 **b** 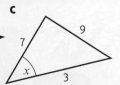 **c**   **d**

   ← **Pure 1 Sections 6.1, 6.2**

You can use vectors to describe relative positions in three dimensions. This allows you to solve geometrical problems in three dimensions and determine properties of 3D solids.

7.1 Vectors

A **vector** has both **magnitude** and **direction**.

You can represent a vector using a **directed line segment**.

This is vector $\overrightarrow{PQ}$. It starts at P and finishes at Q.

This is vector $\overrightarrow{QP}$. It starts at Q and finishes at P.

The direction of the arrow shows the direction of the vector. Small (lower case) letters are also used to represent vectors. In print, the small letter will be in bold type. In writing, you should underline the small letter to show it is a vector: $\underline{a}$ or $\underaccent{\sim}{a}$

- If $\overrightarrow{PQ} = \overrightarrow{RS}$ then the line segments PQ and RS are equal in length and are parallel.

- $\overrightarrow{AB} = -\overrightarrow{BA}$ as the line segment AB is equal in length, parallel and in the opposite direction to BA.

You can add two vectors together using the **triangle law** for vector addition.

- Triangle law for vector addition:
$$\overrightarrow{AB} + \overrightarrow{BC} = \overrightarrow{AC}$$

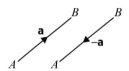

If $\overrightarrow{AB} = \mathbf{a}$, $\overrightarrow{BC} = \mathbf{b}$ and $\overrightarrow{AC} = \mathbf{c}$, then $\mathbf{a} + \mathbf{b} = \mathbf{c}$

Notation The **resultant** is the **vector sum** of two or more vectors.

$$\overrightarrow{AB} + \overrightarrow{BC} + \overrightarrow{CD} = \overrightarrow{AD}$$

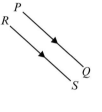

Example **1** **SKILLS** **INTERPRETATION**

The diagram shows vectors **a**, **b** and **c**.

Draw a diagram to illustrate the vector addition
a + b + c

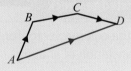

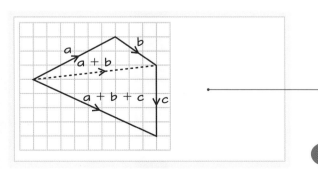

First use the triangle law for **a + b**, then use it again for **(a + b) + c**
The resultant goes from the start of **a** to the end of **c**.

Online Explore vector addition using GeoGebra.

- Subtracting a vector is equivalent to 'adding a negative vector':
  $\mathbf{a} - \mathbf{b} = \mathbf{a} + (-\mathbf{b})$

 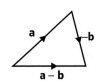

> **Hint** To subtract $\mathbf{b}$, you reverse the direction of $\mathbf{b}$ then add.

If you travel from P to Q, then back from Q to P, you are back where you started, so your **displacement** is zero.

- Adding the vectors $\overrightarrow{PQ}$ and $\overrightarrow{QP}$ gives the zero vector $\mathbf{0}$: $\overrightarrow{PQ} + \overrightarrow{QP} = \mathbf{0}$

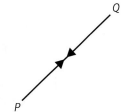

> **Hint** $\overrightarrow{QP} = -\overrightarrow{PQ}$
> So $\overrightarrow{PQ} + \overrightarrow{QP} = \overrightarrow{PQ} - \overrightarrow{PQ} = \mathbf{0}$

You can multiply a vector by a **scalar** (or number).

If the number is positive ($\neq 1$) the new vector has a different length but the **same** direction.

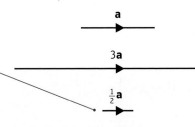

If the number is negative ($\neq -1$) the new vector has a different length and the **opposite** direction.

- Any vector parallel to the vector a may be written as λa, where λ is a non-zero scalar.

> **Notation** Real numbers are examples of **scalars**. They have magnitude but no direction.

Example ②   **SKILLS** ▶ INTERPRETATION

In the diagram, $\overrightarrow{QP} = \mathbf{a}$, $\overrightarrow{QR} = \mathbf{b}$, $\overrightarrow{QS} = \mathbf{c}$ and $\overrightarrow{RT} = \mathbf{d}$

Find in terms of $\mathbf{a}$, $\mathbf{b}$, $\mathbf{c}$ and $\mathbf{d}$

a $\overrightarrow{PS}$ **b** $\overrightarrow{RP}$

c $\overrightarrow{PT}$ **d** $\overrightarrow{TS}$

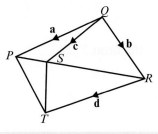

a $\overrightarrow{PS} = \overrightarrow{PQ} + \overrightarrow{QS} = -\mathbf{a} + \mathbf{c}$
$\qquad = \mathbf{c} - \mathbf{a}$

Add vectors using $\triangle PQS$

b $\overrightarrow{RP} = \overrightarrow{RQ} + \overrightarrow{QP} = -\mathbf{b} + \mathbf{a}$
$\qquad = \mathbf{a} - \mathbf{b}$

Add vectors using $\triangle RQP$

c $\overrightarrow{PT} = \overrightarrow{PR} + \overrightarrow{RT} = (\mathbf{b} - \mathbf{a}) + \mathbf{d}$
$\qquad = \mathbf{b} + \mathbf{d} - \mathbf{a}$

Add vectors using $\triangle PRT$
Use $\overrightarrow{PR} = -\overrightarrow{RP} = -(\mathbf{a} - \mathbf{b}) = \mathbf{b} - \mathbf{a}$

d $\overrightarrow{TS} = \overrightarrow{TR} + \overrightarrow{RS} = -\mathbf{d} + (\overrightarrow{RQ} + \overrightarrow{QS})$
$\qquad = -\mathbf{d} + (-\mathbf{b} + \mathbf{c})$
$\qquad = \mathbf{c} - \mathbf{b} - \mathbf{d}$

Add vectors using $\triangle TRS$ and $\triangle RQS$

Example **3** | SKILLS | CRITICAL THINKING

$ABCD$ is a **parallelogram**. $\overrightarrow{AB} = \mathbf{a}$, $\overrightarrow{AD} = \mathbf{b}$ Find $\overrightarrow{AC}$

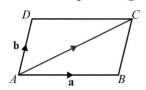

| Notation | This is called the **parallelogram law** for vector addition.

$$\overrightarrow{AC} = \overrightarrow{AB} + \overrightarrow{BC}$$
$$\overrightarrow{BC} = \overrightarrow{AD} = \mathbf{b}$$
$$\text{So } \overrightarrow{AC} = \mathbf{a} + \mathbf{b}$$

Using the triangle law for addition of vectors.

AD and BC are opposite sides of a parallelogram so they are parallel and equal in magnitude.

Example **4** | SKILLS | REASONING

Show that the vectors $6\mathbf{a} + 8\mathbf{b}$ and $9\mathbf{a} + 12\mathbf{b}$ are parallel.

$$9\mathbf{a} + 12\mathbf{b} = \frac{3}{2}(6\mathbf{a} + 8\mathbf{b})$$

$\therefore$ the vectors are parallel.

Here $\lambda = \dfrac{3}{2}$

Example **5** | SKILLS | CREATIVITY

In triangle ABC, $\overrightarrow{AB} = \mathbf{a}$ and $\overrightarrow{AC} = \mathbf{b}$

P is the **midpoint** of AB.

Q divides AC in the ratio $3:2$

Write in terms of $\mathbf{a}$ and $\mathbf{b}$

a $\overrightarrow{BC}$ **b** $\overrightarrow{AP}$ **c** $\overrightarrow{AQ}$ **d** $\overrightarrow{PQ}$

a $\overrightarrow{BC} = \overrightarrow{BA} + \overrightarrow{AC}$
$\qquad = -\overrightarrow{AB} + \overrightarrow{AC}$
$\quad \overrightarrow{BC} = \mathbf{b} - \mathbf{a}$

b $\overrightarrow{AP} = \dfrac{1}{2}\overrightarrow{AB} = \dfrac{1}{2}\mathbf{a}$

c $\overrightarrow{AQ} = \dfrac{3}{5}\overrightarrow{AC} = \dfrac{3}{5}\mathbf{b}$

d $\overrightarrow{PQ} = \overrightarrow{PA} + \overrightarrow{AQ}$
$\qquad = -\overrightarrow{AP} + \overrightarrow{AQ}$
$\qquad = \dfrac{3}{5}\mathbf{b} - \dfrac{1}{2}\mathbf{a}$

$\overrightarrow{BA} = -\overrightarrow{AB}$

$AP = \dfrac{1}{2}AB$ so $\overrightarrow{AP} = \dfrac{1}{2}\mathbf{a}$

| Watch out | AP is the line segment between A and P, whereas $\overrightarrow{AP}$ is the vector from A to P.

Q divides AC in the ratio $3:2$ so $AQ = \dfrac{3}{5}AC$

Going from P to Q is the same as going from P to A, then from A to Q.

Exercise **SKILLS** ▶ INTERPRETATION

1 The diagram shows the vectors **a**, **b**, **c** and **d**.
  Draw a diagram to illustrate these vectors:

  a **a** + **c**   **b** −**b**

  c **c** − **d**   **d** **b** + **c** + **d**

  e 2**c** + 3**d**   **f** **a** − 2**b**

  g **a** + **b** + **c** + **d**

2 *ACGI* is a square, *B* is the midpoint of *AC*, *F* is the midpoint
  of *CG*, *H* is the midpoint of *GI*, *D* is the midpoint of *AI*.
  $\overrightarrow{AB}$ = **b** and $\overrightarrow{AD}$ = **d**. Find, in terms of **b** and **d**

  a $\overrightarrow{AC}$ **b** $\overrightarrow{BE}$ **c** $\overrightarrow{HG}$ **d** $\overrightarrow{DF}$

  e $\overrightarrow{AE}$ **f** $\overrightarrow{DH}$ **g** $\overrightarrow{HB}$ **h** $\overrightarrow{FE}$

  i $\overrightarrow{AH}$ **j** $\overrightarrow{BI}$ **k** $\overrightarrow{EI}$ **l** $\overrightarrow{FB}$

3 *OACB* is a parallelogram. *M*, *Q*, *N* and *P* are
  the midpoints of *OA*, *AC*, *BC* and *OB*
  respectively.

  Vectors **p** and **m** are equal to $\overrightarrow{OP}$ and $\overrightarrow{OM}$
  respectively. Express in terms of **p** and **m**.

  a $\overrightarrow{OA}$ **b** $\overrightarrow{OB}$ **c** $\overrightarrow{BN}$ **d** $\overrightarrow{DQ}$

  e $\overrightarrow{OD}$ **f** $\overrightarrow{MQ}$ **g** $\overrightarrow{OQ}$ **h** $\overrightarrow{AD}$

  i $\overrightarrow{CD}$ **j** $\overrightarrow{AP}$ **k** $\overrightarrow{BM}$ **l** $\overrightarrow{NO}$

4 In the diagram, $\overrightarrow{PQ}$ = **a**, $\overrightarrow{QS}$ = **b**, $\overrightarrow{SR}$ = **c** and $\overrightarrow{PT}$ = **d**
  Find in terms of **a**, **b**, **c** and **d**

  a $\overrightarrow{QT}$ **b** $\overrightarrow{PR}$

  c $\overrightarrow{TS}$ **d** $\overrightarrow{TR}$

5 In the triangle *PQR*, *PQ* = 2**a** and *QR* = 2**b**
  The midpoint of *PR* is *M*. Find, in terms of **a** and **b**

  a $\overrightarrow{PR}$ **b** $\overrightarrow{PM}$ **c** $\overrightarrow{QM}$

(P) 6 *ABCD* is a trapezium with *AB* parallel to *DC* and *DC* = 3*AB*
  M divides *DC* such that *DM* : *MC* = 2 : 1. $\overrightarrow{AB}$ = **a** and $\overrightarrow{BC}$ = **b**
  Find, in terms of **a** and **b**

  a $\overrightarrow{AM}$ **b** $\overrightarrow{BD}$ **c** $\overrightarrow{MB}$ **d** $\overrightarrow{DA}$

Problem-solving

Draw a sketch to show the
information given in the
question.

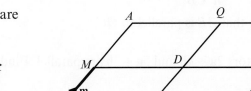

7 $OABC$ is a parallelogram. $\overrightarrow{OA} = \mathbf{a}$ and $\overrightarrow{OC} = \mathbf{b}$

The point P divides OB in the ratio 5:3

Find, in terms of $\mathbf{a}$ and $\mathbf{b}$

a $\overrightarrow{OB}$ **b** $\overrightarrow{OP}$ **c** $\overrightarrow{AP}$

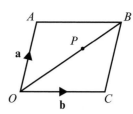

8 State with a reason whether each of these vectors is parallel to the vector $\mathbf{a} - 3\mathbf{b}$.

a $2\mathbf{a} - 6\mathbf{b}$ **b** $4\mathbf{a} - 12\mathbf{b}$ **c** $\mathbf{a} + 3\mathbf{b}$ **d** $3\mathbf{b} - \mathbf{a}$ **e** $9\mathbf{b} - 3\mathbf{a}$ **f** $\frac{1}{2}\mathbf{a} - \frac{2}{3}\mathbf{b}$

(P) **9** In triangle ABC, $\overrightarrow{AB} = \mathbf{a}$ and $\overrightarrow{AC} = \mathbf{b}$

P is the midpoint of AB and Q is the midpoint of AC.

a Write in terms of $\mathbf{a}$ and $\mathbf{b}$

   i $\overrightarrow{BC}$ **ii** $\overrightarrow{AP}$ **iii** $\overrightarrow{AQ}$ **iv** $\overrightarrow{PQ}$

b Show that PQ is parallel to BC.

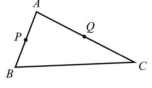

(P) **10** $OABC$ is a **quadrilateral**. $\overrightarrow{OA} = \mathbf{a}$, $\overrightarrow{OC} = 3\mathbf{b}$ and $\overrightarrow{OB} = \mathbf{a} + 2\mathbf{b}$

a Find, in terms of $\mathbf{a}$ and $\mathbf{b}$

   i $\overrightarrow{AB}$ **ii** $\overrightarrow{CB}$

b Show that AB is parallel to OC.

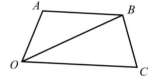

(P) **11** The vectors $2\mathbf{a} + k\mathbf{b}$ and $5\mathbf{a} + 3\mathbf{b}$ are parallel. Find the value of k.

7.2 Representing vectors

A vector can be described by its change in position or **displacement** relative to the x- and y-axes.

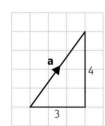

$\mathbf{a} = \begin{pmatrix} 3 \\ 4 \end{pmatrix}$ where 3 is the change in the x-direction and 4 is the change in the y-direction.

This is called **column vector** form.

> **Notation** The top number is the x-component and the bottom number is the y-component.

- To multiply a column vector by a scalar, multiply each component by the scalar: $\lambda \begin{pmatrix} p \\ q \end{pmatrix} = \begin{pmatrix} \lambda p \\ \lambda q \end{pmatrix}$

- To add two column vectors, add the x-components and the y-components: $\begin{pmatrix} p \\ q \end{pmatrix} + \begin{pmatrix} r \\ s \end{pmatrix} = \begin{pmatrix} p + r \\ q + s \end{pmatrix}$

Example **6**   **SKILLS** PROBLEM-SOLVING

$\mathbf{a} = \begin{pmatrix} 2 \\ 6 \end{pmatrix}$ and $\mathbf{b} = \begin{pmatrix} 3 \\ -1 \end{pmatrix}$

Find   **a** $\frac{1}{3}\mathbf{a}$ **b** $\mathbf{a} + \mathbf{b}$ **c** $2\mathbf{a} - 3\mathbf{b}$

a $\frac{1}{3}\mathbf{a} = \begin{pmatrix} \frac{2}{3} \\ 2 \end{pmatrix}$ **Both** of the components are divided by 3.

b $\mathbf{a} + \mathbf{b} = \begin{pmatrix} 2 \\ 6 \end{pmatrix} + \begin{pmatrix} 3 \\ -1 \end{pmatrix} = \begin{pmatrix} 5 \\ 5 \end{pmatrix}$ Add the x-components and the y-components.

c $2\mathbf{a} - 3\mathbf{b} = 2\begin{pmatrix} 2 \\ 6 \end{pmatrix} - 3\begin{pmatrix} 3 \\ -1 \end{pmatrix}$ Multiply each of the vectors by the scalars then subtract the x- and y-components.

$\qquad = \begin{pmatrix} 4 \\ 12 \end{pmatrix} - \begin{pmatrix} 9 \\ -3 \end{pmatrix} = \begin{pmatrix} 4 - 9 \\ 12 + 3 \end{pmatrix} = \begin{pmatrix} -5 \\ 15 \end{pmatrix}$

You can use **unit vectors** to represent vectors in two dimensions.

- A unit vector is a vector of length 1. The unit vectors along the x- and y-axes are usually denoted by **i** and **j** respectively.

  - $\mathbf{i} = \begin{pmatrix} 1 \\ 0 \end{pmatrix}$ $\mathbf{j} = \begin{pmatrix} 0 \\ 1 \end{pmatrix}$

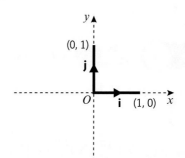

- **You can write any two-dimensional vector in the form $p\mathbf{i} + q\mathbf{j}$**

By the triangle law of addition:

$$\overrightarrow{AC} = \overrightarrow{AB} + \overrightarrow{BC}$$
$$= 5\mathbf{i} + 2\mathbf{j}$$

You can also write this as a **column vector**: $5\mathbf{i} + 2\mathbf{j} = \begin{pmatrix} 5 \\ 2 \end{pmatrix}$

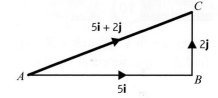

- **For any two-dimensional vector:** $\begin{pmatrix} p \\ q \end{pmatrix} = p\mathbf{i} + q\mathbf{j}$

Example **7**   **SKILLS**   PROBLEM-SOLVING

$\mathbf{a} = 3\mathbf{i} - 4\mathbf{j}, \; \mathbf{b} = 2\mathbf{i} + 7\mathbf{j}$

Find   **a** $\frac{1}{2}\mathbf{a}$ **b** $\mathbf{a} + \mathbf{b}$ **c** $3\mathbf{a} - 2\mathbf{b}$

a $\frac{1}{2}\mathbf{a} = \frac{1}{2}(3\mathbf{i} - 4\mathbf{j}) = 1.5\mathbf{i} - 2\mathbf{j}$ Divide the **i** component and the **j** component by 2.

b $\mathbf{a} + \mathbf{b} = 3\mathbf{i} - 4\mathbf{j} + 2\mathbf{i} + 7\mathbf{j}$
$\qquad = (3 + 2)\mathbf{i} + (-4 + 7)\mathbf{j} = 5\mathbf{i} + 3\mathbf{j}$ Add the **i** components and the **j** components.

c $3\mathbf{a} - 2\mathbf{b} = 3(3\mathbf{i} - 4\mathbf{j}) - 2(2\mathbf{i} + 7\mathbf{j})$
$\qquad = 9\mathbf{i} - 12\mathbf{j} - (4\mathbf{i} + 14\mathbf{j})$
$\qquad = (9 - 4)\mathbf{i} + (-12 - 14)\mathbf{j}$ Multiply each of the vectors by the scalar then subtract the **i** and **j** components.
$\qquad = 5\mathbf{i} - 26\mathbf{j}$

Example **8**   **SKILLS**   INTERPRETATION

a Draw a diagram to represent the vector $-3\mathbf{i} + \mathbf{j}$

b Write this as a column vector.

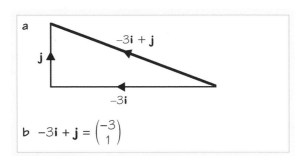

3 units in the direction of the unit vector $-\mathbf{i}$ and 1 unit in the direction of the unit vector $\mathbf{j}$.

b $-3\mathbf{i} + \mathbf{j} = \begin{pmatrix} -3 \\ 1 \end{pmatrix}$

Example **9**   **SKILLS**   PROBLEM-SOLVING

Given that $\mathbf{a} = 2\mathbf{i} + 5\mathbf{j}$, $\mathbf{b} = 12\mathbf{i} - 10\mathbf{j}$ and $\mathbf{c} = -3\mathbf{i} + 9\mathbf{j}$, find $\mathbf{a} + \mathbf{b} + \mathbf{c}$, using column vector notation in your working.

$$\mathbf{a} + \mathbf{b} + \mathbf{c} = \begin{pmatrix} 2 \\ 5 \end{pmatrix} + \begin{pmatrix} 12 \\ -10 \end{pmatrix} + \begin{pmatrix} -3 \\ 9 \end{pmatrix} = \begin{pmatrix} 11 \\ 4 \end{pmatrix}$$

Add the numbers in the top line to get 11 (the x-component), and the bottom line to get 4 (the y-component). This is $11\mathbf{i} + 4\mathbf{j}$

Example **10**   **SKILLS**   PROBLEM-SOLVING

Given $\mathbf{a} = 5\mathbf{i} + 2\mathbf{j}$ and $\mathbf{b} = 3\mathbf{i} - 4\mathbf{j}$, find $2\mathbf{a} - \mathbf{b}$ in terms of $\mathbf{i}$ and $\mathbf{j}$.

Online   Explore this solution as a vector diagram on a coordinate grid using GeoGebra.

$$2\mathbf{a} = 2\begin{pmatrix} 5 \\ 2 \end{pmatrix} = \begin{pmatrix} 10 \\ 4 \end{pmatrix}$$

To find the column vector for vector $2\mathbf{a}$ multiply the $\mathbf{i}$ and $\mathbf{j}$ components of vector $\mathbf{a}$ by 2.

$$2\mathbf{a} - \mathbf{b} = \begin{pmatrix} 10 \\ 4 \end{pmatrix} - \begin{pmatrix} 3 \\ -4 \end{pmatrix} = \begin{pmatrix} 10 - 3 \\ 4 - (-4) \end{pmatrix} = \begin{pmatrix} 7 \\ 8 \end{pmatrix}$$

To find the column vector for $2\mathbf{a} - \mathbf{b}$ subtract the components of vector $\mathbf{b}$ from those of vector $2\mathbf{a}$.

$$2\mathbf{a} - \mathbf{b} = 7\mathbf{i} + 8\mathbf{j}$$

Remember to give your answer in terms of $\mathbf{i}$ and $\mathbf{j}$.

Exercise **7B**   **SKILLS**   INTERPRETATION

1 These vectors are drawn on a grid of unit squares. Express the vectors $\mathbf{v}_1$, $\mathbf{v}_2$, $\mathbf{v}_3$, $\mathbf{v}_4$, $\mathbf{v}_5$ and $\mathbf{v}_6$ in $\mathbf{i}$, $\mathbf{j}$ notation and column vector form.

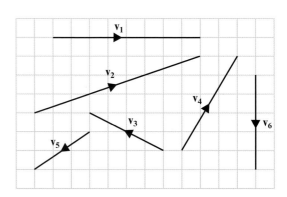

2 Given that **a** = 2**i** + 3**j** and **b** = 4**i** − **j**, find these vectors in terms of **i** and **j**.

    a 4**a**   **b** $\frac{1}{2}$**a**   **c** −**b**   **d** 2**b** + **a**

    e 3**a** − 2**b**   **f** **b** − 3**a**   **g** 4**b** − **a**   **h** 2**a** − 3**b**

3 Given that **a** = $\begin{pmatrix} 9 \\ 7 \end{pmatrix}$, **b** = $\begin{pmatrix} 11 \\ -3 \end{pmatrix}$ and **c** = $\begin{pmatrix} -8 \\ -1 \end{pmatrix}$ find

    a 5**a**   **b** $-\frac{1}{2}$**c**   **c** **a** + **b** + **c**   **d** 2**a** − **b** + **c**

    e 2**b** + 2**c** − 3**a**   **f** $\frac{1}{2}$**a** + $\frac{1}{2}$**b**

(P) 4 Given that **a** = 2**i** + 5**j** and **b** = 3**i** − **j**, find

    a λ if **a** + λ**b** is parallel to the vector **i**   **b** μ if μ**a** + **b** is parallel to the vector **j**

(P) 5 Given that **c** = 3**i** + 4**j** and **d** = **i** − 2**j** find

    a λ if **c** + λ**d** is parallel to **i** + **j**   **b** μ if μ**c** + **d** is parallel to **i** + 3**j**

    c s if **c** − s**d** is parallel to 2**i** + **j**   **d** t if **d** − t**c** is parallel to −2**i** + 3**j**

(E) 6 In triangle ABC, $\overrightarrow{AB}$ = 4**i** + 3**j** and $\overrightarrow{AC}$ = 5**i** + 2**j**

    Find $\overrightarrow{BC}$.

                                 (2 marks)

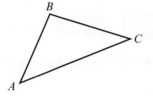

(P) 7 $OABC$ is a parallelogram.

    P divides AC in the ratio $3:2$. $\overrightarrow{OA}$ = 2**i** + 4**j**, $\overrightarrow{OC}$ = 7**i**

    Find in **i**, **j** format and column vector format:

    a $\overrightarrow{AC}$ **b** $\overrightarrow{AP}$ **c** $\overrightarrow{OP}$

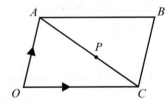

(E/P) 8 **a** = $\begin{pmatrix} j \\ 3 \end{pmatrix}$, **b** = $\begin{pmatrix} 10 \\ k \end{pmatrix}$, **c** = $\begin{pmatrix} 2 \\ 5 \end{pmatrix}$

    Given that **b** − 2**a** = **c**, find the values of j and k.

                            (2 marks)

> **Problem-solving**
>
> You can consider **b** − 2**a** = **c** as two linear equations. One for the x-components and one for the y-components.

(E/P) 9 **a** = $\begin{pmatrix} p \\ -q \end{pmatrix}$, **b** = $\begin{pmatrix} q \\ p \end{pmatrix}$, **c** = $\begin{pmatrix} 7 \\ 4 \end{pmatrix}$

    Given that **a** + 2**b** = **c**, find the values of p and q.   **(2 marks)**

(E/P) 10 The resultant of the vectors **a** = 3**i** − 2**j** and **b** = p**i** − 2p**j** is parallel to the vector **c** = 2**i** − 3**j**

    Find

    a the value of p **(4 marks)**

    b the resultant of vectors **a** and **b**.   **(1 mark)**

7.3 Magnitude and direction

You can use Pythagoras' theorem to calculate the **magnitude** of a vector.

- For the vector $\mathbf{a} = x\mathbf{i} + y\mathbf{j} = \begin{pmatrix} x \\ y \end{pmatrix}$

  the magnitude of the vector is given by:

  $$|\mathbf{a}| = \sqrt{x^2 + y^2}$$

Notation You use straight lines on either side of the vector:

$$|\mathbf{a}| = |x\mathbf{i} + y\mathbf{j}| = \left| \begin{pmatrix} x \\ y \end{pmatrix} \right|$$

You need to be able to find a **unit vector** in the direction of a given vector.

- A unit vector in the direction of $\mathbf{a}$ is $\dfrac{\mathbf{a}}{|\mathbf{a}|}$

If $|\mathbf{a}| = 5$ then a unit vector in the direction of $\mathbf{a}$ is $\dfrac{\mathbf{a}}{5}$

Notation A unit vector is any vector with magnitude 1.
A unit vector in the direction of $\mathbf{a}$ is sometimes written as $\hat{\mathbf{a}}$.

Example 11   **SKILLS**   CRITICAL THINKING

Given that $\mathbf{a} = 3\mathbf{i} + 4\mathbf{j}$ and $\mathbf{b} = -2\mathbf{i} - 4\mathbf{j}$

a find $|\mathbf{a}|$

b find a unit vector in the direction of $\mathbf{a}$

c find the exact value of $|2\mathbf{a} + \mathbf{b}|$

Online Explore the magnitude of a vector using GeoGebra.

a $\mathbf{a} = \begin{pmatrix} 3 \\ 4 \end{pmatrix}$

$|\mathbf{a}| = \sqrt{3^2 + 4^2}$

$|\mathbf{a}| = \sqrt{25} = 5$

It is often quicker and easier to convert from $\mathbf{i}, \mathbf{j}$ form to column vector form for calculations.

Using Pythagoras.

b a unit vector is $\dfrac{\mathbf{a}}{|\mathbf{a}|} = \dfrac{3\mathbf{i} + 4\mathbf{j}}{5}$

$= \dfrac{1}{5}(3\mathbf{i} + 4\mathbf{j})$ or $\begin{pmatrix} 0.6 \\ 0.8 \end{pmatrix}$

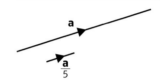

c $2\mathbf{a} + \mathbf{b} = 2\begin{pmatrix} 3 \\ 4 \end{pmatrix} + \begin{pmatrix} -2 \\ -4 \end{pmatrix} = \begin{pmatrix} 6 - 2 \\ 8 - 4 \end{pmatrix} = \begin{pmatrix} 4 \\ 4 \end{pmatrix}$

$|2\mathbf{a} + \mathbf{b}| = \sqrt{4^2 + 4^2} = \sqrt{32} = 4\sqrt{2}$

Unless specified in the question it is acceptable to give your answer in $\mathbf{i}, \mathbf{j}$ form or column vector form.

You need to give an exact answer, so leave your answer in surd form:

$$\sqrt{32} = \sqrt{16 \times 2} = 4\sqrt{2}$$

← Pure 1 Section 1.5

You can define a vector by giving its magnitude, and the angle between the vector and one of the coordinate axes. This is called **magnitude-direction form**.

Example (12)   **SKILLS**   INTERPRETATION

Find the angle between the vector 4**i** + 5**j** and the positive x-axis.

> This might be referred to as the angle between the vector and **i**.

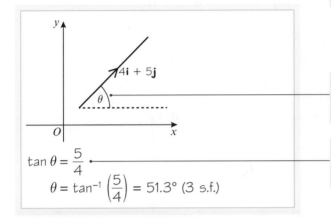

> Identify the angle that you need to find. A diagram always helps.

$$\tan \theta = \frac{5}{4}$$

> You have a **right-angled** triangle with base 4 units and height 5 units, so use trigonometry.

$$\theta = \tan^{-1}\left(\frac{5}{4}\right) = 51.3° \ (3 \text{ s.f.})$$

Example (13)   **SKILLS**   ANALYSIS

Vector **a** has magnitude 10 and makes an angle of 30° with **j**.
Find **a** in **i**, **j** and column vector format.

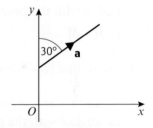

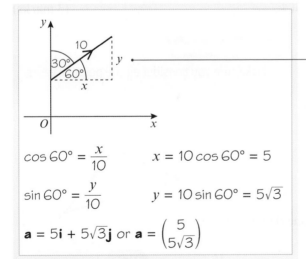

> Use trigonometry to find the lengths of the x- and y-components for vector **a**.

$$\cos 60° = \frac{x}{10} \qquad x = 10 \cos 60° = 5$$

$$\sin 60° = \frac{y}{10} \qquad y = 10 \sin 60° = 5\sqrt{3}$$

Watch out   The direction of a vector can be given relative to either the positive x-axis (the **i** direction) or the positive y-axis (or the **j** direction).

$$\mathbf{a} = 5\mathbf{i} + 5\sqrt{3}\mathbf{j} \ \text{ or } \ \mathbf{a} = \begin{pmatrix} 5 \\ 5\sqrt{3} \end{pmatrix}$$

Exercise (7C)   **SKILLS**   PROBLEM-SOLVING

1 Find the magnitude of each of these vectors.

     a 3**i** + 4**j**   **b** 6**i** − 8**j**   **c** 5**i** + 12**j**   **d** 2**i** + 4**j**

     e 3**i** − 5**j**   **f** 4**i** + 7**j**   **g** −3**i** + 5**j**   **h** −4**i** − **j**

2   $\mathbf{a} = 2\mathbf{i} + 3\mathbf{j}$, $\mathbf{b} = 3\mathbf{i} - 4\mathbf{j}$ and $\mathbf{c} = 5\mathbf{i} - \mathbf{j}$. Find the exact value of the magnitude of

    a $\mathbf{a} + \mathbf{b}$ **b** $2\mathbf{a} - \mathbf{c}$ **c** $3\mathbf{b} - 2\mathbf{c}$

3   For each of the following vectors, find the unit vector in the same direction.

    a $\mathbf{a} = 4\mathbf{i} + 3\mathbf{j}$ **b** $\mathbf{b} = 5\mathbf{i} - 12\mathbf{j}$ **c** $\mathbf{c} = -7\mathbf{i} + 24\mathbf{j}$ **d** $\mathbf{d} = \mathbf{i} - 3\mathbf{j}$

4   Find the angle that each of these vectors makes with the positive x-axis.

    a $3\mathbf{i} + 4\mathbf{j}$ **b** $6\mathbf{i} - 8\mathbf{j}$ **c** $5\mathbf{i} + 12\mathbf{j}$ **d** $2\mathbf{i} + 4\mathbf{j}$

5   Find the angle that each of these vectors makes with $\mathbf{j}$.

    a $3\mathbf{i} - 5\mathbf{j}$ **b** $4\mathbf{i} + 7\mathbf{j}$ **c** $-3\mathbf{i} + 5\mathbf{j}$ **d** $-4\mathbf{i} - \mathbf{j}$

6   Write these vectors in $\mathbf{i}$, $\mathbf{j}$ and column vector form.

    a 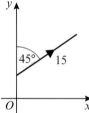   **b** 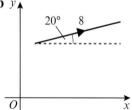   **c**   **d**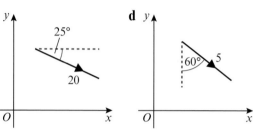

7   Draw a sketch for each vector and work out the exact value of its magnitude and the angle it makes with the positive x-axis to one decimal place.

    a $3\mathbf{i} + 4\mathbf{j}$ **b** $2\mathbf{i} - \mathbf{j}$ **c** $-5\mathbf{i} + 2\mathbf{j}$

(E/P)   8   Given that $|2\mathbf{i} - k\mathbf{j}| = 2\sqrt{10}$, find the exact value of k.   **(3 marks)**

(E/P)   9   Vector $\mathbf{a} = p\mathbf{i} + q\mathbf{j}$ has magnitude 10 and makes an angle θ with the positive x-axis where $\sin \theta = \frac{3}{5}$. Find the possible values of p and q.

(4 marks)

> **Problem-solving**
>
> Make sure you consider all the possible cases.

10   In triangle ABC, $\overrightarrow{AB} = 4\mathbf{i} + 3\mathbf{j}$, $\overrightarrow{AC} = 6\mathbf{i} - 4\mathbf{j}$

    a   Find the angle between $\overrightarrow{AB}$ and $\mathbf{i}$.

    b   Find the angle between $\overrightarrow{AC}$ and $\mathbf{i}$.

    c   Hence find the size of $\angle BAC$, in **degrees**, to one decimal place.

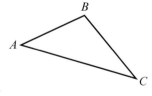

(E/P)   11   In triangle PQR, $\overrightarrow{PQ} = 4\mathbf{i} + \mathbf{j}$, $\overrightarrow{PR} = 6\mathbf{i} - 8\mathbf{j}$

    a   Find the size of $\angle QPR$, in degrees, to one decimal place.   **(5 marks)**

    b   Find the area of triangle PQR.   **(2 marks)**

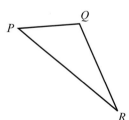

> **Hint**   The area of a triangle is $\frac{1}{2}ab \sin \theta$
>
>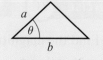
>
> ← **Pure 1 Section 6.2**

7.4   Vectors in 3D

You can use 3D vectors to describe position and displacement relative to the x-, y- and z-axes. You can represent 3D vectors as column vectors or using the unit vectors **i**, **j** and **k**.

- The unit vectors along the x-, y- and z-axes are denoted by **i**, **j** and **k** respectively.

$$\mathbf{i} = \begin{pmatrix} 1 \\ 0 \\ 0 \end{pmatrix} \qquad \mathbf{j} = \begin{pmatrix} 0 \\ 1 \\ 0 \end{pmatrix} \qquad \mathbf{k} = \begin{pmatrix} 0 \\ 0 \\ 1 \end{pmatrix}$$

> **Links**   3D vectors obey all the same addition and scalar multiplication rules as 2D vectors.
>   ← **Pure 4 Section 7.1**

- For any 3D vector $p\mathbf{i} + q\mathbf{j} + r\mathbf{k} = \begin{pmatrix} p \\ q \\ r \end{pmatrix}$

Example 14   SKILLS   PROBLEM-SOLVING

Consider the points $A(1, 5, -2)$ and $B(0, -3, 7)$

a Find the **position vectors** of A and B in **ijk** notation.

b Find the vector $\overrightarrow{AB}$ as a column vector.

---

a $\overrightarrow{OA} = \mathbf{i} + 5\mathbf{j} - 2\mathbf{k}$, $\overrightarrow{OB} = -3\mathbf{j} + 7\mathbf{k}$

b $\overrightarrow{AB} = \overrightarrow{OB} - \overrightarrow{OA}$

$$= \begin{pmatrix} 0 \\ -3 \\ 7 \end{pmatrix} - \begin{pmatrix} 1 \\ 5 \\ -2 \end{pmatrix} = \begin{pmatrix} -1 \\ -8 \\ 9 \end{pmatrix}$$

> The position vector of a point is the vector from the origin to that point.

> $\overrightarrow{OB}$ has no component in the **i** direction. You could write it as $0\mathbf{i} - 3\mathbf{j} + 7\mathbf{k}$

> When adding and subtracting vectors it is often easier to write them as column vectors.

Example 15   SKILLS   REASONING

The vectors **a** and **b** are given as $\mathbf{a} = \begin{pmatrix} 2 \\ -3 \\ 5 \end{pmatrix}$ and $\mathbf{b} = \begin{pmatrix} 4 \\ -2 \\ 0 \end{pmatrix}$

> **Online**   Perform calculations on 3D vectors using your calculator.

a Find

   i $4\mathbf{a} + \mathbf{b}$ **ii** $2\mathbf{a} - 3\mathbf{b}$

b State with a reason whether each of these vectors is parallel to $4\mathbf{i} - 5\mathbf{k}$

---

a   **i** $4\mathbf{a} + \mathbf{b} = 4\begin{pmatrix} 2 \\ -3 \\ 5 \end{pmatrix} + \begin{pmatrix} 4 \\ -2 \\ 0 \end{pmatrix}$

$$= \begin{pmatrix} 8 \\ -12 \\ 20 \end{pmatrix} + \begin{pmatrix} 4 \\ -2 \\ 0 \end{pmatrix}$$

$$= \begin{pmatrix} 12 \\ -14 \\ 20 \end{pmatrix}$$

> Use the rules for scalar multiplication and addition of vectors:
>
> $$\lambda \begin{pmatrix} p \\ q \\ r \end{pmatrix} = \begin{pmatrix} \lambda p \\ \lambda q \\ \lambda r \end{pmatrix} \text{ and } \begin{pmatrix} p \\ q \\ r \end{pmatrix} + \begin{pmatrix} u \\ v \\ w \end{pmatrix} = \begin{pmatrix} p + u \\ q + v \\ r + w \end{pmatrix}$$

ii $2\mathbf{a} - 3\mathbf{b} = 2\begin{pmatrix} 2 \\ -3 \\ 5 \end{pmatrix} - 3\begin{pmatrix} 4 \\ -2 \\ 0 \end{pmatrix}$

$= \begin{pmatrix} 4 \\ -6 \\ 10 \end{pmatrix} - \begin{pmatrix} 12 \\ -6 \\ 0 \end{pmatrix}$

$= \begin{pmatrix} -8 \\ 0 \\ 10 \end{pmatrix}$

b   i $4\mathbf{a} + \mathbf{b} = \begin{pmatrix} 12 \\ -14 \\ 20 \end{pmatrix} = 3\begin{pmatrix} 4 \\ -\frac{14}{3} \\ \frac{20}{3} \end{pmatrix}$

which is not a multiple of $\begin{pmatrix} 4 \\ 0 \\ -5 \end{pmatrix}$

> Two vectors are parallel if one is a multiple of the other. Make the x-components the same and compare the y- and z-components with $4\mathbf{i} - 5\mathbf{k}$

$4\mathbf{a} + \mathbf{b}$ is not parallel to $4\mathbf{i} - 5\mathbf{k}$

ii $2\mathbf{a} - 3\mathbf{b} = \begin{pmatrix} -8 \\ 0 \\ 10 \end{pmatrix} = -2\begin{pmatrix} 4 \\ 0 \\ -5 \end{pmatrix}$

which is a multiple of $\begin{pmatrix} 4 \\ 0 \\ -5 \end{pmatrix}$

$2\mathbf{a} - 3\mathbf{b}$ is parallel to $4\mathbf{i} - 5\mathbf{k}$

> **Watch out** $4\mathbf{i} - 5\mathbf{k} = 4\mathbf{i} + 0\mathbf{j} - 5\mathbf{k}$
> Make sure you include a 0 in the $\mathbf{j}$-component of the column vector.

Example **16**   **SKILLS**   PROBLEM-SOLVING

Find the magnitude of $\mathbf{a} = 2\mathbf{i} - \mathbf{j} + 4\mathbf{k}$ and hence find $\hat{\mathbf{a}}$, the unit vector in the direction of $\mathbf{a}$.

The magnitude of $\mathbf{a}$ is given by

$|\mathbf{a}| = \sqrt{2^2 + (-1)^2 + 4^2}$

$= \sqrt{21}$

$\hat{\mathbf{a}} = \dfrac{\mathbf{a}}{|\mathbf{a}|} = \dfrac{1}{\sqrt{21}}(2\mathbf{i} - \mathbf{j} + 4\mathbf{k})$

> Use Pythagoras' theorem.

> You could also write this as $\dfrac{2}{\sqrt{21}}\mathbf{i} - \dfrac{1}{\sqrt{21}}\mathbf{j} + \dfrac{4}{\sqrt{21}}\mathbf{k}$

> **Online**   Check your answer using the vector functions on your calculator.

You can find the angle between a given vector and any of the coordinate axes by considering the appropriate right-angled triangle.

- If the vector $\mathbf{a} = x\mathbf{i} + y\mathbf{j} + z\mathbf{k}$ makes an angle θ_x with the positive x-axis then $\cos \theta_x = \dfrac{x}{|\mathbf{a}|}$ and similarly for the angles θ_y and θ_z

> **Hint**   This rule also works with vectors in two dimensions.

Example 17   **SKILLS**   INTERPRETATION

Find the angles that the vector $\mathbf{a} = 2\mathbf{i} - 3\mathbf{j} - \mathbf{k}$ makes with each of the positive coordinate axes to 1 d.p.

$|\mathbf{a}| = \sqrt{2^2 + (-3)^3 + (-1)^2} = \sqrt{4 + 9 + 1} = \sqrt{14}$ •——— First find $|\mathbf{a}|$ since you will be using it three times.

$\cos\theta_x = \dfrac{x}{|\mathbf{a}|} = \dfrac{2}{\sqrt{14}} = 0.5345...$

$\theta_x = 57.7°$ (1 d.p.)

Write down at least 4 d.p., or use the answer button on your calculator to enter the exact value.

$\cos\theta_y = \dfrac{y}{|\mathbf{a}|} = \dfrac{-3}{\sqrt{14}} = -0.8017...$

$\theta_y = 143.3°$ (1 d.p.)

The formula also works with negative components. The y-component is negative, so the vector makes an obtuse angle with the positive y-axis.

$\cos\theta_z = \dfrac{z}{|\mathbf{a}|} = \dfrac{-1}{\sqrt{14}} = -0.2672...$

$\theta_z = 105.5°$ (1 d.p.)

Example 18   **SKILLS**   ANALYSIS

The points A and B have position vectors $4\mathbf{i} + 2\mathbf{j} + 7\mathbf{k}$ and $3\mathbf{i} + 4\mathbf{j} - \mathbf{k}$ relative to a fixed origin, O. Find $\overrightarrow{AB}$ and show that $\triangle OAB$ is **isosceles**.

$\overrightarrow{OA} = \mathbf{a} = \begin{pmatrix} 4 \\ 2 \\ 7 \end{pmatrix}$, $\overrightarrow{OB} = \mathbf{b} = \begin{pmatrix} 3 \\ 4 \\ -1 \end{pmatrix}$ •——— Write down the position vectors of A and B.

$\overrightarrow{AB} = \mathbf{b} - \mathbf{a} = \begin{pmatrix} 3 \\ 4 \\ -1 \end{pmatrix} - \begin{pmatrix} 4 \\ 2 \\ 7 \end{pmatrix} = \begin{pmatrix} -1 \\ 2 \\ -8 \end{pmatrix}$ •——— Use $\overrightarrow{AB} = \mathbf{b} - \mathbf{a}$

$|\overrightarrow{AB}| = \sqrt{(-1)^2 + 2^2 + (-8)^2} = \sqrt{69}$ •——— This is the length of the line segment AB.

$|\overrightarrow{OA}| = \sqrt{4^2 + 2^2 + 7^2} = \sqrt{69}$

$|\overrightarrow{OB}| = \sqrt{3^2 + 4^2 + (-1)^2} = \sqrt{26}$

Find the lengths of the other sides OA and OB of $\triangle OAB$

So $\triangle OAB$ is isosceles, with $AB = OA$

Online Explore the solution to this example visually in 3D using GeoGebra.

Exercise 7D   **SKILLS**   CRITICAL THINKING

1 The vectors $\mathbf{a}$ and $\mathbf{b}$ are defined by $\mathbf{a} = \begin{pmatrix} 1 \\ 2 \\ -4 \end{pmatrix}$ and $\mathbf{b} = \begin{pmatrix} 4 \\ -3 \\ 5 \end{pmatrix}$

  a Find

    i $\mathbf{a} - \mathbf{b}$ **ii** $-\mathbf{a} + 3\mathbf{b}$

  b State with a reason whether each of these vectors is parallel to $6\mathbf{i} - 10\mathbf{j} + 18\mathbf{k}$

2 The vectors $\mathbf{a}$ and $\mathbf{b}$ are defined by $\mathbf{a} = \begin{pmatrix} 3 \\ 2 \\ -1 \end{pmatrix}$ and $\mathbf{b} = \begin{pmatrix} -3 \\ -2 \\ 4 \end{pmatrix}$

  Show that the vector $3\mathbf{a} + 2\mathbf{b}$ is parallel to $6\mathbf{i} + 4\mathbf{j} + 10\mathbf{k}$

(P)   **3** The vectors **a** and **b** are defined by $\mathbf{a} = \begin{pmatrix} 1 \\ 2 \\ -4 \end{pmatrix}$ and $\mathbf{b} = \begin{pmatrix} p \\ q \\ r \end{pmatrix}$

Given that $\mathbf{a} + 2\mathbf{b} = 5\mathbf{i} + 4\mathbf{j}$ find the values of p, q and r.

4 Find the magnitude of

   a $3\mathbf{i} + 5\mathbf{j} + \mathbf{k}$ **b** $4\mathbf{i} - 2\mathbf{k}$ **c** $\mathbf{i} + \mathbf{j} - \mathbf{k}$

   d $5\mathbf{i} - 9\mathbf{j} - 8\mathbf{k}$ **e** $\mathbf{i} + 5\mathbf{j} - 7\mathbf{k}$

5 Given that $\mathbf{p} = \begin{pmatrix} 5 \\ 0 \\ 2 \end{pmatrix}$, $\mathbf{q} = \begin{pmatrix} 2 \\ 1 \\ -3 \end{pmatrix}$ and $\mathbf{r} = \begin{pmatrix} 7 \\ -4 \\ 2 \end{pmatrix}$, find in column vector form:

   a $\mathbf{p} + \mathbf{q}$ **b** $\mathbf{q} - \mathbf{r}$ **c** $\mathbf{p} + \mathbf{q} + \mathbf{r}$

   d $3\mathbf{p} - \mathbf{r}$ **e** $\mathbf{p} - 2\mathbf{q} + \mathbf{r}$

6 The position vector of the point A is $2\mathbf{i} - 7\mathbf{j} + 3\mathbf{k}$ and $\overrightarrow{AB} = 5\mathbf{i} + 4\mathbf{j} - \mathbf{k}$. Find the position vector of the point B.

(P)   **7** Given that $\mathbf{a} = t\mathbf{i} + 2\mathbf{j} + 3\mathbf{k}$ and that $|\mathbf{a}| = 7$ find the possible values of t.

(P)   **8** Given that $\mathbf{a} = 5t\mathbf{i} + 2t\mathbf{j} + t\mathbf{k}$ and that $|\mathbf{a}| = 3\sqrt{10}$ find the possible values of t.

9 The points A, B and C have coordinates $(2, 1, 4)$, $(3, -2, 4)$ and $(-1, 2, 2)$.

   a   Find, in terms of $\mathbf{i}$, $\mathbf{j}$ and $\mathbf{k}$

     i   the position vectors of A, B and C

     ii $\overrightarrow{AC}$

   b   Find the exact value of

     i $\left| \overrightarrow{AC} \right|$

     ii $\left| \overrightarrow{OC} \right|$

10 P is the point $(3, 0, 7)$ and Q is the point $(-1, 3, -5)$. Find

   a   the vector $\overrightarrow{PQ}$

   b   the distance between P and Q

   c   the unit vector in the direction of $\overrightarrow{PQ}$.

11 $\overrightarrow{OA}$ is the vector $4\mathbf{i} - \mathbf{j} - 2\mathbf{k}$ and $\overrightarrow{OB}$ is the vector $-2\mathbf{i} + 3\mathbf{j} + \mathbf{k}$. Find

   a   the vector $\overrightarrow{AB}$

   b   the distance between A and B

   c   the unit vector in the direction of $\overrightarrow{AB}$.

12 Find the unit vector in the direction of each of the following vectors.

a $\mathbf{p} = \begin{pmatrix} 3 \\ -4 \\ -2 \end{pmatrix}$ **b** $\mathbf{q} = \begin{pmatrix} \sqrt{2} \\ -4 \\ -\sqrt{7} \end{pmatrix}$ **c** $\mathbf{r} = \begin{pmatrix} \sqrt{5} \\ -2\sqrt{2} \\ -\sqrt{3} \end{pmatrix}$

(E/P) **13** The points A, B and C have position vectors $\begin{pmatrix} 8 \\ -7 \\ 4 \end{pmatrix}$, $\begin{pmatrix} 8 \\ -3 \\ 3 \end{pmatrix}$ and $\begin{pmatrix} 12 \\ -6 \\ 3 \end{pmatrix}$ respectively.

     a Find the vectors $\overrightarrow{AB}$, $\overrightarrow{AC}$ and $\overrightarrow{BC}$ **(3 marks)**

     b Find $|\overrightarrow{AB}|$, $|\overrightarrow{AC}|$ and $|\overrightarrow{BC}|$ giving your answers in exact form.   **(6 marks)**

     c Describe triangle ABC.   **(1 mark)**

(E) **14** A is the point $(3, 4, 8)$, B is the point $(1, -2, 5)$ and C is the point $(7, -5, 7)$.

     a Find the vectors $\overrightarrow{AB}$, $\overrightarrow{AC}$ and $\overrightarrow{BC}$ **(3 marks)**

     b Hence find the lengths of the sides of triangle ABC.   **(6 marks)**

     c Given that angle $ABC = 90°$ find the size of angle BAC.   **(2 marks)**

15 For each of the given vectors,

     a $-\mathbf{i} + 7\mathbf{j} + \mathbf{k}$ **b** $\begin{pmatrix} 3 \\ 4 \\ 7 \end{pmatrix}$ **c** $\begin{pmatrix} 2 \\ 0 \\ -3 \end{pmatrix}$

     find the angle made by the vector with:

     i the positive x-axis   **ii** the positive y-axis   **iii** the positive z-axis

(P) **16** A **scalene** triangle has the coordinates $(2, 0, 0)$, $(5, 0, 0)$ and $(4, 2, 3)$.
     Work out the area of the triangle.

(E/P) **17** The diagram shows the triangle PQR.

     Given that $\overrightarrow{PQ} = 3\mathbf{i} - \mathbf{j} + 2\mathbf{k}$ and

     $\overrightarrow{QR} = -2\mathbf{i} + 4\mathbf{j} + 3\mathbf{k}$, show that
     $\angle PQR = 78.5°$ to 1 d.p.

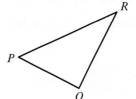

                                                             (5 marks)

Challenge

Find the **acute angle** that the vector $\mathbf{a} = -2\mathbf{i} + 6\mathbf{j} - 3\mathbf{k}$ makes with the xy-**plane**.
Give your answer to 1 d.p.

7.5   Solving geometric problems in two dimensions

You need to be able to use vectors to solve geometric problems and to find the position vector of a point that divides a line segment in a given ratio.

■ If the point P divides the line segment AB in the ratio $\lambda : \mu$, then

$$\overrightarrow{OP} = \overrightarrow{OA} + \frac{\lambda}{\lambda + \mu}\overrightarrow{AB}$$

$$= \overrightarrow{OA} + \frac{\lambda}{\lambda + \mu}(\overrightarrow{OB} - \overrightarrow{OA})$$

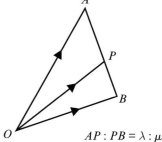

$AP : PB = \lambda : \mu$

Example (19)   **SKILLS**   INTERPRETATION

In the diagram the points A and B have position vectors **a** and **b** respectively (referred to the origin O). The point P divides AB in the ratio $1 : 2$.

Find the position vector of P.

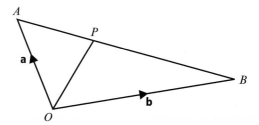

$$\overrightarrow{OP} = \overrightarrow{OA} + \frac{1}{3}\overrightarrow{AB}$$

$$= \overrightarrow{OA} + \frac{1}{3}(\overrightarrow{OB} - \overrightarrow{OA})$$

$$= \frac{2}{3}\overrightarrow{OA} + \frac{1}{3}\overrightarrow{OB}$$

$$= \frac{2}{3}\mathbf{a} + \frac{1}{3}\mathbf{b}$$

There are 3 parts in the ratio in total, so P is $\frac{1}{3}$ of the way along the line segment AB.

Rewrite $\overrightarrow{AB}$ in terms of the position vectors for A and B.

Give your final answer in terms of **a** and **b**.

You can solve geometric problems by comparing coefficients on both sides of an equation:

■ If **a** and **b** are two **non-parallel** vectors and $p\mathbf{a} + q\mathbf{b} = r\mathbf{a} + s\mathbf{b}$ then $p = r$ and $q = s$

Example (20)   **SKILLS**   REASONING

$OABC$ is a parallelogram. P is the point where the diagonals OB and AC intersect.

The vectors **a** and **c** are equal to $\overrightarrow{OA}$ and $\overrightarrow{OC}$ respectively.

Prove that the diagonals bisect each other.

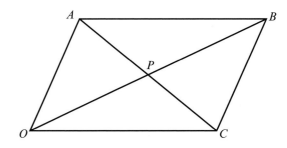

If the diagonals bisect each other, then P must be the midpoint of OB and the midpoint of AC.

From the diagram,

$$\overrightarrow{OB} = \overrightarrow{OC} + \overrightarrow{CB} = \mathbf{c} + \mathbf{a}$$

and $\overrightarrow{AC} = \overrightarrow{AO} + \overrightarrow{OC}$

$$= -\overrightarrow{OA} + \overrightarrow{OC} = -\mathbf{a} + \mathbf{c}$$

P lies on OB $\Rightarrow$ $\overrightarrow{OP} = \lambda(\mathbf{c} + \mathbf{a})$

P lies on AC $\Rightarrow$ $\overrightarrow{OP} = \overrightarrow{OA} + \overrightarrow{AP}$

$$= \mathbf{a} + \mu(-\mathbf{a} + \mathbf{c})$$

$\Rightarrow$ $\lambda(\mathbf{c} + \mathbf{a}) = \mathbf{a} + \mu(-\mathbf{a} + \mathbf{c})$

$\Rightarrow$ $\lambda = 1 - \mu$ and $\lambda = \mu$

$\Rightarrow$ $\lambda = \mu = \dfrac{1}{2}$, so P is the midpoint of both

diagonals, so the diagonals bisect each other.

Online Use technology to show that **diagonals** of a parallelogram **bisect** each other.

Express $\overrightarrow{OB}$ and $\overrightarrow{AC}$ in terms of $\mathbf{a}$ and $\mathbf{c}$.

Use the fact that P lies on both diagonals to find two different routes from O to P, giving two different forms of $\overrightarrow{OP}$

The two expressions for $\overrightarrow{OP}$ must be equal.

Form and solve a pair of simultaneous equations by equating the coefficients of $\mathbf{a}$ and $\mathbf{c}$.

If P is halfway along the line segment then it must be the midpoint.

Example 21 **SKILLS** PROBLEM-SOLVING

In triangle ABC, $\overrightarrow{AB} = 3\mathbf{i} - 2\mathbf{j}$ and $\overrightarrow{AC} = \mathbf{i} - 5\mathbf{j}$

Find the exact size of $\angle BAC$ in degrees.

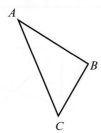

Problem-solving

Work out what information you would need to find the angle. You could:

- find the lengths of all three sides then use the cosine rule
- convert $\overrightarrow{AB}$ and $\overrightarrow{AC}$ to magnitude-direction form

The working here shows the first method.

$$\overrightarrow{BC} = \overrightarrow{AC} - \overrightarrow{AB} = \begin{pmatrix} 1 \\ -5 \end{pmatrix} - \begin{pmatrix} 3 \\ -2 \end{pmatrix} = \begin{pmatrix} -2 \\ -3 \end{pmatrix}$$

$$|\overrightarrow{AB}| = \sqrt{3^2 + (-2)^2} = \sqrt{13}$$

$$|\overrightarrow{AC}| = \sqrt{1^2 + (-5)^2} = \sqrt{26}$$

$$|\overrightarrow{BC}| = \sqrt{(-2)^2 + (-3)^2} = \sqrt{13}$$

$$\cos \angle BAC = \frac{|\overrightarrow{AB}|^2 + |\overrightarrow{AC}|^2 - |\overrightarrow{BC}|^2}{2 \times |\overrightarrow{AB}| \times |\overrightarrow{AC}|}$$

$$= \frac{13 + 26 - 13}{2 \times \sqrt{13} \times \sqrt{26}} = \frac{26}{26\sqrt{2}} = \frac{1}{\sqrt{2}}$$

$$\angle BAC = \cos^{-1}\left(\frac{1}{\sqrt{2}}\right) = 45°$$

Use the triangle law to find $\overrightarrow{BC}$

Leave your answers in surd form.

$$\cos A = \frac{b^2 + c^2 - a^2}{2bc}$$

Online Check your answer by entering the vectors directly into your calculator.

Exercise **7E**   **SKILLS**   REASONING/ARGUMENTATION

(P) **1** In the diagram, $\overrightarrow{WX} = \mathbf{a}$, $\overrightarrow{WY} = \mathbf{b}$ and
$\overrightarrow{WZ} = \mathbf{c}$. It is given that $\overrightarrow{XY} = \overrightarrow{YZ}$
Prove that $\mathbf{a} + \mathbf{c} = 2\mathbf{b}$

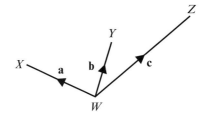

(P) **2** *OAB* is a triangle. *P*, *Q* and *R* are the midpoints
of *OA*, *AB* and *OB* respectively.

OP and *OR* are equal to **p** and **r** respectively.

a Find   **i** $\overrightarrow{OB}$ **ii** $\overrightarrow{PQ}$

b Hence prove that triangle *PAQ* is similar to triangle *OAB*.

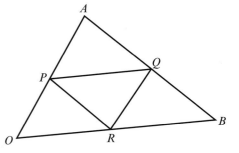

(P) **3** *OAB* is a triangle. $\overrightarrow{OA} = \mathbf{a}$ and $\overrightarrow{OB} = \mathbf{b}$
The point *M* divides *OA* in the ratio 2 : 1
MN is parallel to *OB*.

a Express the vector $\overrightarrow{ON}$ in terms of **a** and **b**.

b Show that $AN : NB = 1 : 2$

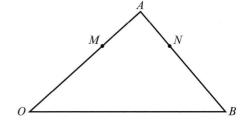

(P) **4** *OABC* is a square. *M* is the midpoint of *OA*, and *Q* divides *BC*
in the ratio 1 : 3
AC and *MQ* meet at *P*.

a If $\overrightarrow{OA} = \mathbf{a}$ and $\overrightarrow{OC} = \mathbf{c}$, express $\overrightarrow{OP}$ in terms of **a** and **c**.

b Show that *P* divides *AC* in the ratio 2 : 3

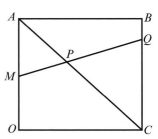

5 In triangle *ABC* the position vectors of the vertices *A*, *B* and *C* are $\begin{pmatrix} 5 \\ 8 \end{pmatrix}$, $\begin{pmatrix} 4 \\ 3 \end{pmatrix}$ and $\begin{pmatrix} 7 \\ 6 \end{pmatrix}$. Find

a $|\overrightarrow{AB}|$ **b** $|\overrightarrow{AC}|$ **c** $|\overrightarrow{BC}|$

d the size of $\angle BAC$, $\angle ABC$ and $\angle ACB$ to the nearest degree.

P **6** OPQ is a triangle.

$2\overrightarrow{PR} = \overrightarrow{RQ}$ and $3\overrightarrow{OR} = \overrightarrow{OS}$

$\overrightarrow{OP} = \mathbf{a}$ and $\overrightarrow{OQ} = \mathbf{b}$.

a Show that $\overrightarrow{OS} = 2\mathbf{a} + \mathbf{b}$

b Point T is added to the diagram such that $\overrightarrow{OT} = -\mathbf{b}$

Prove that points T, P and S lie on a straight line.

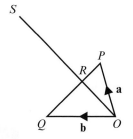

> **Problem-solving**
>
> To show that T, P and S lie on the same straight line you need to show that any **two** of the vectors $\overrightarrow{TP}$, $\overrightarrow{TS}$ or $\overrightarrow{PS}$ are parallel.

Challenge

$OPQR$ is a parallelogram.

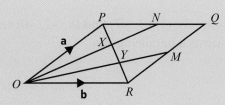

N is the midpoint of PQ and M is the midpoint of QR. $\overrightarrow{OP} = \mathbf{a}$ and $\overrightarrow{OR} = \mathbf{b}$
The lines ON and OM intersect the diagonal PR at points X and Y respectively.

a Explain why $\overrightarrow{PX} = -j\mathbf{a} + j\mathbf{b}$, where j is a constant.

b Show that $\overrightarrow{PX} = (k-1)\mathbf{a} + \frac{1}{2}k\mathbf{b}$, where k is a constant.

c Explain why the values of j and k must satisfy these simultaneous equations:

$k - 1 = -j$

$\frac{1}{2}k = j$

d Hence find the values of j and k.

e Deduce that the lines ON and OM divide the diagonal PR into 3 equal parts.

7.6 **Solving geometric problems in three dimensions**

You need to be able to solve geometric problems involving vectors in three dimensions.

Example **22**   **SKILLS** CRITICAL THINKING

A, B, C and D are the points $(2, -5, -8)$, $(1, -7, -3)$, $(0, 15, -10)$ and $(2, 19, -20)$ respectively.

a Find $\overrightarrow{AB}$ and $\overrightarrow{DC}$, giving your answers in the form $p\mathbf{i} + q\mathbf{j} + r\mathbf{k}$

b Show that the lines AB and DC are parallel and that $\overrightarrow{DC} = 2\overrightarrow{AB}$

c Hence describe the quadrilateral $ABCD$.

a $\overrightarrow{AB} = \overrightarrow{OB} - \overrightarrow{OA}$

$= (\mathbf{i} - 7\mathbf{j} - 3\mathbf{k}) - (2\mathbf{i} - 5\mathbf{j} - 8\mathbf{k})$

$= -\mathbf{i} - 2\mathbf{j} + 5\mathbf{k}$

$\overrightarrow{DC} = \overrightarrow{OC} - \overrightarrow{OD}$

$= (15\mathbf{j} - 10\mathbf{k}) - (2\mathbf{i} + 19\mathbf{j} - 20\mathbf{k})$

$= -2\mathbf{i} - 4\mathbf{j} + 10\mathbf{k}$

> **Watch out** AB refers to the **line segment** between A and B (or its length), whereas $\overrightarrow{AB}$ refers to the **vector** from A to B. Note that $AB = BA$ but $\overrightarrow{AB} \neq \overrightarrow{BA}$

b $2\overrightarrow{AB} = 2(-\mathbf{i} - 2\mathbf{j} + 5\mathbf{k})$
$= -2\mathbf{i} - 4\mathbf{j} + 10\mathbf{k} = \overrightarrow{DC}$

So AB is parallel to DC and half as long.

c There are two unequal parallel sides, so $ABCD$ is a trapezium.

Problem-solving

If you can't work out what shape it is, draw a sketch showing AB and DC.

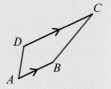

Online Explore the solution to this example visually in 3D using GeoGebra.

Example 23 **SKILLS** CRITICAL THINKING

P, Q and R are the points $(4, -9, -3)$, $(7, -7, -7)$ and $(8, -2, -0)$ respectively. Find the coordinates of the point S so that $PQRS$ forms a parallelogram.

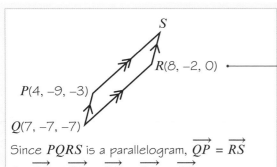

Since $PQRS$ is a parallelogram, $\overrightarrow{QP} = \overrightarrow{RS}$

So $\overrightarrow{OS} = \overrightarrow{OR} + \overrightarrow{RS} = \overrightarrow{OR} + \overrightarrow{QP}$

$\overrightarrow{OR} = \begin{pmatrix} 8 \\ -2 \\ 0 \end{pmatrix}$ and

$\overrightarrow{QP} = \overrightarrow{OP} - \overrightarrow{OQ} = \begin{pmatrix} 4 \\ -9 \\ -3 \end{pmatrix} - \begin{pmatrix} 7 \\ -7 \\ -7 \end{pmatrix} = \begin{pmatrix} -3 \\ -2 \\ 4 \end{pmatrix}$

So $\overrightarrow{OS} = \begin{pmatrix} 8 \\ -2 \\ 0 \end{pmatrix} + \begin{pmatrix} -3 \\ -2 \\ 4 \end{pmatrix} = \begin{pmatrix} 5 \\ -4 \\ 4 \end{pmatrix}$

which means that S is the point $(5, -4, 4)$

Draw a sketch. The vertices in a 2D shape are given in order, either **clockwise** or **anticlockwise**. It doesn't matter if the positions on the sketch don't correspond to the real positions in 3D – it is still a helpful way to visualise the problem.

You could also go from O to S via P:

$\overrightarrow{OS} = \overrightarrow{OP} + \overrightarrow{PS}$
$= \overrightarrow{OP} + \overrightarrow{QR}$

In two dimensions you saw that if **a** and **b** are two non-parallel vectors and $p\mathbf{a} + q\mathbf{b} = r\mathbf{a} + s\mathbf{b}$ then $p = r$ and $q = s$. In other words, in two dimensions with two vectors you can **compare coefficients** on both sides of an equation. In three dimensions you have to extend this rule:

Notation **Coplanar vectors** are vectors which are in the same plane.
Non-coplanar vectors are vectors which are **not** in the same plane.

- If **a**, **b** and **c** are vectors in three dimensions which do not all lie on the same plane then you can compare their coefficients on both sides of an equation.

In particular, since the vectors **i**, **j** and **k** are non-**coplanar**, if $p\mathbf{i} + q\mathbf{j} + r\mathbf{k} = u\mathbf{i} + v\mathbf{j} + w\mathbf{k}$ then $p = u$, $q = v$ and $r = w$

Example **24**   **SKILLS**   PROBLEM-SOLVING

Given that $3\mathbf{i} + (p + 2)\mathbf{j} + 120\mathbf{k} = p\mathbf{i} - q\mathbf{j} + 4pqr\mathbf{k}$, find the values of p, q and r.

Comparing coefficients of $\mathbf{i}$ gives $p = 3$.

Comparing coefficients of $\mathbf{j}$ gives $p + 2 = -q$
so $q = -(3 + 2) = -5$.

Comparing coefficients of $\mathbf{k}$ gives
$120 = 4pqr$ so $r = \dfrac{120}{4 \times 3 \times (-5)} = -2$

Since $\mathbf{i}$, $\mathbf{j}$ and $\mathbf{k}$ do not lie in the same plane you can compare coefficients.

When comparing coefficients like this just write the coefficients. For example, write $3 = p$, not $3\mathbf{i} = p\mathbf{i}$

Example **25**   **SKILLS**   ANALYSIS

The diagram shows a cuboid whose vertices are O, A, B, C, D, E, F and G. Vectors $\mathbf{a}$, $\mathbf{b}$ and $\mathbf{c}$ are the position vectors of the vertices A, B and C respectively. Prove that the diagonals OE and BG bisect each other.

Hint Bisect means 'cut into two equal parts'. In this case you need to prove that **both** diagonals are bisected.

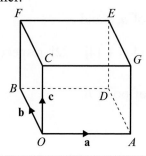

Problem-solving

If there is a point of intersection, H, it must lie on both diagonals. You can reach H directly from O (travelling along OE), or by first travelling to B then travelling along BG. Use this to write two expressions for $\overrightarrow{OH}$

Suppose there is a point of intersection, H, of OE and BG.

$\overrightarrow{OH} = r\overrightarrow{OE}$ for some scalar r.

But $\overrightarrow{OH} = \overrightarrow{OB} + \overrightarrow{BH}$ and $\overrightarrow{BH} = s\overrightarrow{BG}$
for some scalar s, so $\overrightarrow{OH} = \overrightarrow{OB} + s\overrightarrow{BG}$

So $r\overrightarrow{OE} = \overrightarrow{OB} + s\overrightarrow{BG}$ **(1)**

Now $\overrightarrow{OE} = \overrightarrow{OA} + \overrightarrow{AD} + \overrightarrow{DE} = \mathbf{a} + \mathbf{b} + \mathbf{c}$,
$\overrightarrow{OB} = \mathbf{b}$ and $\overrightarrow{BG} = \overrightarrow{OG} - \overrightarrow{OB} = \mathbf{a} + \mathbf{c} - \mathbf{b}$

So **(1)** becomes $r(\mathbf{a} + \mathbf{b} + \mathbf{c}) = \mathbf{b} + s(\mathbf{a} + \mathbf{c} - \mathbf{b})$

Comparing coefficients in $\mathbf{a}$ and $\mathbf{b}$ gives $r = s$
and $r = 1 - s$

Solving simultaneously gives $r = s = \dfrac{1}{2}$

These solutions also satisfy the coefficients of $\mathbf{c}$ so the lines do intersect at H.

$OH = \dfrac{1}{2}OE$ so H bisects OE.

$BH = \dfrac{1}{2}BG$ so H bisects BG, as required.

H lies on the line OE, so $\overrightarrow{OH}$ must be some scalar multiple of $\overrightarrow{OE}$

Use the fact that H lies on both diagonals to find two different expression for $\overrightarrow{OH}$. You can equate these expressions and compare coefficients.

$\mathbf{a}$, $\mathbf{b}$ and $\mathbf{c}$ are three non-coplanar vectors so you can compare coefficients.

In order for the lines to intersect, the values of r and s must satisfy equation (1) completely:
$\dfrac{1}{2}(\mathbf{a} + \mathbf{b} + \mathbf{c}) = \mathbf{b} + \dfrac{1}{2}(\mathbf{a} + \mathbf{c} - \mathbf{b})$
The coefficients of $\mathbf{a}$, $\mathbf{b}$ and $\mathbf{c}$ all match so both ways of writing the vector OH are identical.

Vector proofs such as this one often avoid any coordinate **geometry**, which tends to be messy and complicated, especially in three dimensions.

Exercise 7F   **SKILLS**   ADAPTIVE LEARNING

(P) 1 The points A, B and C have position vectors $\begin{pmatrix} 1 \\ -4 \\ 8 \end{pmatrix}$, $\begin{pmatrix} 4 \\ 4 \\ 7 \end{pmatrix}$ and $\begin{pmatrix} 10 \\ 0 \\ 30 \end{pmatrix}$ relative to a fixed origin, O.

   a Show that:

     i $|\overrightarrow{OA}| = |\overrightarrow{OB}|$ **ii** $|\overrightarrow{AC}| = |\overrightarrow{BC}|$

   b Hence describe the quadrilateral $OACB$.

(P) 2 The points A, B and C have coordinates $(2, 1, 5)$, $(4, 4, 3)$ and $(2, 7, 5)$ respectively.

   a Show that triangle ABC is isosceles.

   b Find the area of triangle ABC.

   c Find a point D such that $ABCD$ is a parallelogram.

(P) 3 The points A, B, C and D have coordinates $(7, 12, -1)$, $(11, 2, -9)$, $(14, -14, 3)$ and $(8, 1, 15)$ respectively.

   a Show that AB and CD are parallel, and find the ratio $AB : CD$ in its simplest form.

   b Hence describe the quadrilateral $ABCD$.

(P) 4 Given that $(3a + b)\mathbf{i} + \mathbf{j} + ac\mathbf{k} = 7\mathbf{i} - b\mathbf{j} + 4\mathbf{k}$, find the values of a, b and c.

(P) 5 The points A and B have position vectors $10\mathbf{i} - 23\mathbf{j} + 10\mathbf{k}$ and $p\mathbf{i} + 14\mathbf{j} - 22\mathbf{k}$ respectively, relative to a fixed origin O, where p is a constant.

   Given that $\triangle OAB$ is isosceles, find **three** possible positions of point B.

(E/P) 6 The diagram shows a triangle ABC.

   Given that $\overrightarrow{AB} = 7\mathbf{i} - \mathbf{j} + 2\mathbf{k}$ and $\overrightarrow{BC} = -\mathbf{i} + 5\mathbf{k}$

   a find the area of triangle ABC.   **(7 marks)**

   The point D is such that $\overrightarrow{AD} = 3\overrightarrow{AB}$,

   and the point E is such that $\overrightarrow{AE} = 3\overrightarrow{AC}$

   b Find the area of triangle ADE.   **(2 marks)**

(P) 7 A **parallelepiped** is a three-dimensional figure formed by six parallelograms. The diagram shows a parallelepiped with vertices O, A, B, C, D, E, F, and G. $\mathbf{a}$, $\mathbf{b}$ and $\mathbf{c}$ are the vectors $\overrightarrow{OA}$, $\overrightarrow{OB}$ and $\overrightarrow{OC}$ respectively. Prove that the diagonals OF and AG bisect each other.

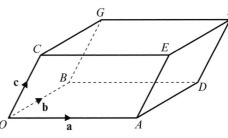

(P) 8 The diagram shows a cuboid whose vertices are O, A, B, C, D, E, F and G. $\mathbf{a}$, $\mathbf{b}$ and $\mathbf{c}$ are the position vectors of the vertices A, B and C respectively. The point M lies on OE such that $OM : ME = 3 : 1$. The straight line AP passes through point M. Given that $AM : MP = 3 : 1$ prove that P lies on the line EF and find the ratio $FP : PE$

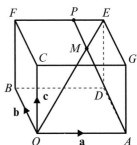

Challenge

1   **a**, **b** and **c** are the vectors $\begin{pmatrix} 1 \\ 0 \\ 4 \end{pmatrix}$, $\begin{pmatrix} 2 \\ 0 \\ -3 \end{pmatrix}$ and $\begin{pmatrix} -5 \\ 3 \\ 1 \end{pmatrix}$ respectively. Find scalars

   p, q and r such that $p\mathbf{a} + q\mathbf{b} + r\mathbf{c} = \begin{pmatrix} 28 \\ -12 \\ -4 \end{pmatrix}$

2   The diagram shows a cuboid with vertices O, A, B, C, D, E, F and G. M is the midpoint of FE and N is the midpoint of AG.

   a, **b** and **c** are the position vectors of the vertices A, B and C respectively.

   Prove that the lines OM and BN **trisect** the diagonal AF.

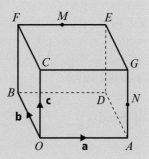

Hint   Trisect means divide into three equal parts.

7.7   Position vectors

You need to be able to use vectors to describe the position of a point in two dimensions.

Position vectors are vectors giving the position of a point, relative to a fixed origin.

The position vector of a point A is the vector $\overrightarrow{OA}$, where O is the origin.

If $\overrightarrow{OA} = a\mathbf{i} + b\mathbf{j}$ then the position vector of A is $\begin{pmatrix} a \\ b \end{pmatrix}$

- In general, a point P with coordinates (p, q) has a position vector
  $$\overrightarrow{OP} = p\mathbf{i} + q\mathbf{j} = \begin{pmatrix} p \\ q \end{pmatrix}$$

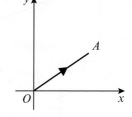

- $\overrightarrow{AB} = \overrightarrow{OB} - \overrightarrow{OA}$, where $\overrightarrow{OA}$ and $\overrightarrow{OB}$ are the position vectors of A and B respectively.

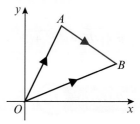

Example 26   **SKILLS**   PROBLEM-SOLVING

The points A and B in the diagram have coordinates $(3, 4)$ and $(11, 2)$ respectively.
Find, in terms of **i** and **j**

a the position vector of A **b** the position vector of B

c the vector $\overrightarrow{AB}$

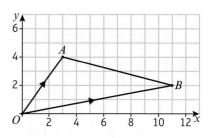

a $\overrightarrow{OA} = 3\mathbf{i} + 4\mathbf{j}$ ————————— In column vector form this is $\begin{pmatrix} 3 \\ 4 \end{pmatrix}$

b $\overrightarrow{OB} = 11\mathbf{i} + 2\mathbf{j}$ ————————— In column vector form this is $\begin{pmatrix} 11 \\ 2 \end{pmatrix}$

c $\overrightarrow{AB} = \overrightarrow{OB} - \overrightarrow{OA}$
$= (11\mathbf{i} + 2\mathbf{j}) - (3\mathbf{i} + 4\mathbf{j}) = 8\mathbf{i} - 2\mathbf{j}$ ————— In column vector form this is $\begin{pmatrix} 8 \\ -2 \end{pmatrix}$

Example **27** **SKILLS** PROBLEM-SOLVING

$\overrightarrow{OA} = 5\mathbf{i} - 2\mathbf{j}$ and $\overrightarrow{AB} = 3\mathbf{i} + 4\mathbf{j}$. Find

a the position vector of B

b the exact value of $|\overrightarrow{OB}|$ in simplified surd form.

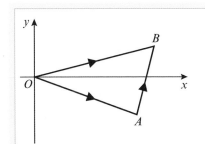

a $\overrightarrow{OA} = \begin{pmatrix} 5 \\ -2 \end{pmatrix}$ and $\overrightarrow{AB} = \begin{pmatrix} 3 \\ 4 \end{pmatrix}$ ————— It is usually quicker to use column vector form for calculations.

$\overrightarrow{OB} = \overrightarrow{OA} + \overrightarrow{AB} = \begin{pmatrix} 5 \\ -2 \end{pmatrix} + \begin{pmatrix} 3 \\ 4 \end{pmatrix} = \begin{pmatrix} 8 \\ 2 \end{pmatrix}$ ————— In $\mathbf{i}, \mathbf{j}$ form the answer is $8\mathbf{i} + 2\mathbf{j}$

b $|\overrightarrow{OB}| = \sqrt{8^2 + 2^2} = \sqrt{64 + 4} = \sqrt{68} = 2\sqrt{17}$ ————— $\sqrt{68} = \sqrt{4 \times 17} = 2\sqrt{17}$ in simplified surd form.

Exercise **7G** **SKILLS** CRITICAL THINKING

1 The points A, B and C have coordinates $(3, -1)$, $(4, 5)$ and $(-2, 6)$ respectively, and O is the origin.

   a Find, in terms of $\mathbf{i}$ and $\mathbf{j}$

     i the position vectors of A, B and C **ii** $\overrightarrow{AB}$ **iii** $\overrightarrow{AC}$

   b Find, in surd form

     i $|\overrightarrow{OC}|$ **ii** $|\overrightarrow{AB}|$ **iii** $|\overrightarrow{AC}|$

2 $\overrightarrow{OP} = 4\mathbf{i} - 3\mathbf{j}$, $\overrightarrow{OQ} = 3\mathbf{i} + 2\mathbf{j}$

   a Find $\overrightarrow{PQ}$

   b Find, in surd form   **i** $|\overrightarrow{OP}|$ **ii** $|\overrightarrow{OQ}|$ **iii** $|\overrightarrow{PQ}|$

3 $\overrightarrow{OQ} = 4\mathbf{i} - 3\mathbf{j}$, $\overrightarrow{PQ} = 5\mathbf{i} + 6\mathbf{j}$

   a Find $\overrightarrow{OP}$

   b Find, in surd form: **i** $|\overrightarrow{OP}|$ **ii** $|\overrightarrow{OQ}|$ **iii** $|\overrightarrow{PQ}|$

(P) **4** *OABCDE* is a regular **hexagon**. The points *A* and *B* have position vectors **a** and **b** respectively, where *O* is the origin.

   Find, in terms of **a** and **b**, the position vectors of

   a *C*   **b** *D*   **c** *E*

(P) **5** The position vectors of 3 vertices of a parallelogram

   are $\begin{pmatrix} 4 \\ 2 \end{pmatrix}$, $\begin{pmatrix} 3 \\ 5 \end{pmatrix}$ and $\begin{pmatrix} 8 \\ 6 \end{pmatrix}$

   Find the possible position vectors of the fourth vertex.

> **Problem-solving**
>
> Use a sketch to check that you have considered all the possible positions for the fourth vertex.

(E) **6** Given that the point *A* has position vector $4\mathbf{i} - 5\mathbf{j}$ and the point *B* has position vector $6\mathbf{i} + 3\mathbf{j}$

   a find the vector $\overrightarrow{AB}$ **(2 marks)**

   b find $|\overrightarrow{AB}|$ giving your answer as a simplified surd.   **(2 marks)**

(E/P) **7** The point *A* lies on the circle with equation $x^2 + y^2 = 9$

   Given that $\overrightarrow{OA} = 2k\mathbf{i} + k\mathbf{j}$ find the exact value of *k*.   **(3 marks)**

> **Challenge**
>
> The point *B* lies on the line with equation $2y = 12 - 3x$
>
> Given that $|\overrightarrow{OB}| = \sqrt{13}$ find possible expressions for $\overrightarrow{OB}$ in the form $p\mathbf{i} + q\mathbf{j}$

7.8 3D coordinates

Cartesian coordinate axes in three dimensions are usually called *x*-, *y*- and *z*-axes, each being at right angles to each of the others.

The coordinates of a point in three dimensions are written as (x, y, z).

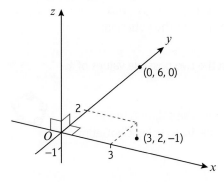

> **Hint**   To visualise this, think of the *x*- and *y*-axes being drawn on a flat surface and the *z*-axis sticking up from the surface.

You can use Pythagoras' theorem in 3D to find distances on a 3D coordinate grid.

■ **The distance from the origin to the point (x, y, z) is $\sqrt{x^2 + y^2 + z^2}$**

Example (31) **SKILLS** PROBLEM-SOLVING

Find the distance from the origin to the point $P(4, -7, -1)$

$OP = \sqrt{4^2 + (-7)^2 + (-1)^2}$

$\quad = \sqrt{16 + 49 + 1}$

$\quad = \sqrt{66}$

Substitute the values of x, y and z into the formula. You don't need to give units with distances on a coordinate grid.

You can also use Pythagoras' theorem to find the distance between two points.

■ **The distance between the points (x_1, y_1, z_1) and (x_2, y_2, z_2) is**

$$\sqrt{(x_1 - x_2)^2 + (y_1 - y_2)^2 + (z_1 - z_2)^2}$$

Example (32) **SKILLS** PROBLEM-SOLVING

Find the distance between the points $A(1, 3, 4)$ and $B(8, 6, -5)$, giving your answer to 1 d.p.

$AB = \sqrt{(1 - 8)^2 + (3 - 6)^2 + (4 - (-5))^2}$

$\quad = \sqrt{(-7)^2 + (-3)^2 + 9^2}$

$\quad = \sqrt{49 + 9 + 81}$

$\quad = \sqrt{139} = 11.8$ (1 d.p.)

Be careful with the signs – use brackets when you substitute.

Example (33) **SKILLS** CRITICAL THINKING

The coordinates of A and B are $(5, 0, 3)$ and $(4, 2, k)$ respectively.

Given that the distance from A to B is 3 units, find the possible values of k.

$AB = \sqrt{(5 - 4)^2 + (0 - 2)^2 + (3 - k)^2} = 3$

$\quad\quad \sqrt{1 + 4 + (9 - 6k + k^2)} = 3$

$\quad\quad\quad 1 + 4 + 9 - 6k + k^2 = 9$

$\quad\quad\quad\quad k^2 - 6k + 5 = 0$

$\quad\quad\quad\quad (k - 5)(k - 1) = 0$

$k = 1$ or $k = 5$

Problem-solving

Use Pythagoras' theorem to form a quadratic equation in k.

Square both sides of the equation.

Solve to find the two possible values of k.

Online Explore the solution to this example visually in 3D using GeoGebra.

Exercise **7H** SKILLS PROBLEM-SOLVING

1 Find the distance from the origin to the point $P(2, 8, -4)$

2 Find the distance from the origin to the point $P(7, 7, 7)$

3 Find the distance between A and B when they have the following coordinates:

  a $A(3, 0, 5)$ and $B(1, -1, 8)$

  b $A(8, 11, 8)$ and $B(-3, 1, 6)$

  c $A(3, 5, -2)$ and $B(3, 10, 3)$

  d $A(-1, -2, 5)$ and $B(4, -1, 3)$

(P) **4** The coordinates of A and B are $(7, -1, 2)$ and $(k, 0, 4)$ respectively.
Given that the distance from A to B is 3 units, find the possible values of k.

(P) **5** The coordinates of A and B are $(5, 3, -8)$ and $(1, k, -3)$ respectively.
Given that the distance from A to B is $3\sqrt{10}$ units, find the possible values of k.

Challenge

  a The points $A(1, 3, -2)$, $B(1, 3, 4)$ and $C(7, -3, 4)$ are three vertices of a solid cube.
    Write down the coordinates of the remaining five vertices.

  An ant walks from A to C along the surface of the cube.

  b Determine the length of the shortest possible route the ant can take.

7.9 Equation of a line in three dimensions

You need to know how to write the equation of a straight line in vector form.

Suppose a straight line passes through a given point A, with position vector **a**, and is parallel to the given vector **b**. Only one such line is possible. Let R be an arbitrary point on the line, with position vector **r**.

Since $\overrightarrow{AR}$ is parallel to **b**, $\overrightarrow{AR} = \lambda\mathbf{b}$, where λ is a scalar.

The vector **b** is called the **direction vector** of the line.

So the position vector **r** can be written as $\mathbf{a} + \lambda\mathbf{b}$

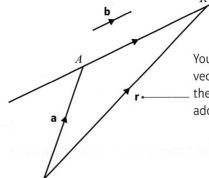

You can find the position vector of any point R on the line by using vector addition ($\triangle OAR$):

$$\mathbf{r} = \mathbf{a} + \overrightarrow{AR}$$

- A vector equation of a straight line passing through the point A with position vector **a**, and parallel to the vector **b** is

    $\mathbf{r} = \mathbf{a} + \lambda\mathbf{b}$

where λ is a scalar parameter.

Notation **r** is the position vector of a general point on the line. Scalar parameters in vector equations are often given Greek letters such as λ (lambda) and μ (mu).

By taking different values of the parameter λ, you can find the position vectors of different points that lie on the straight line.

Online Explore the vector equation of a line using GeoGebra.

Example 31 **SKILLS** PROBLEM-SOLVING

Find a vector equation of the straight line which passes through the point A, with position vector $3\mathbf{i} - 5\mathbf{j} + 4\mathbf{k}$ and is parallel to the vector $7\mathbf{i} - 3\mathbf{k}$

Here $\mathbf{a} = \begin{pmatrix} 3 \\ -5 \\ 4 \end{pmatrix}$ and $\mathbf{b} = \begin{pmatrix} 7 \\ 0 \\ -3 \end{pmatrix}$ ——— $\mathbf{b}$ is the direction vector.

An equation of the line is

$$\mathbf{r} = \begin{pmatrix} 3 \\ -5 \\ 4 \end{pmatrix} + \lambda \begin{pmatrix} 7 \\ 0 \\ -3 \end{pmatrix}$$

or $\mathbf{r} = (3\mathbf{i} - 5\mathbf{j} + 4\mathbf{k}) + \lambda(7\mathbf{i} - 3\mathbf{k})$

or $\mathbf{r} = (3 + 7\lambda)\mathbf{i} + (-5)\mathbf{j} + (4 - 3\lambda)\mathbf{k}$

or $\mathbf{r} = \begin{pmatrix} 3 + 7\lambda \\ -5 \\ 4 - 3\lambda \end{pmatrix}$

You sometimes need to show the separate x, y, z components in terms of λ.

You can represent a 3D vector using column notation, $\begin{pmatrix} x \\ y \\ z \end{pmatrix}$, or using **ijk** notation, $x\mathbf{i} + y\mathbf{j} + z\mathbf{k}$

← **Pure 4 Section 7.4**

Now suppose a straight line passes through two given points C and D, with position vectors $\mathbf{c}$ and $\mathbf{d}$ respectively. Again, only one such line is possible.

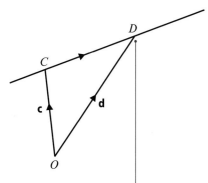

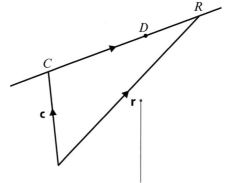

You can use $\overrightarrow{CD}$ as a direction vector for the line:
$\overrightarrow{CD} = \mathbf{d} - \mathbf{c}$

You can now use one of the two given points and the direction vector to form an equation for the straight line.

- **A vector equation of a straight line passing through the points C and D, with position vectors $\mathbf{c}$ and $\mathbf{d}$ respectively, is**

$$\mathbf{r} = \mathbf{c} + \lambda(\mathbf{d} - \mathbf{c})$$

where λ is a scalar parameter.

Notation You can use any point on the straight line as the initial point in the vector equation. An alternative vector equation for this line would be $\mathbf{r} = \mathbf{d} + \lambda(\mathbf{d} - \mathbf{c})$

Example (32)   **SKILLS**   PROBLEM-SOLVING

Find a vector equation of the straight line which passes through the points A and B, with coordinates $(4, 5, -1)$ and $(6, 3, 2)$ respectively.

$$\mathbf{a} = \begin{pmatrix} 4 \\ 5 \\ -1 \end{pmatrix} \quad \mathbf{b} = \begin{pmatrix} 6 \\ 3 \\ 2 \end{pmatrix}$$ Write down the position vectors of A and B.

$$\mathbf{b} - \mathbf{a} = \begin{pmatrix} 6 \\ 3 \\ 2 \end{pmatrix} - \begin{pmatrix} 4 \\ 5 \\ -1 \end{pmatrix} = \begin{pmatrix} 2 \\ -2 \\ 3 \end{pmatrix}$$ Find a direction vector for the line.

Use one of the given points to form the equation.

$$\mathbf{r} = \begin{pmatrix} 4 \\ 5 \\ -1 \end{pmatrix} + t\begin{pmatrix} 2 \\ -2 \\ 3 \end{pmatrix}$$

or $\mathbf{r} = (4\mathbf{i} + 5\mathbf{j} - \mathbf{k}) + t(2\mathbf{i} - 2\mathbf{j} + 3\mathbf{k})$

or $\mathbf{r} = (4 + 2t)\mathbf{i} + (5 - 2t)\mathbf{j} + (-1 + 3t)\mathbf{k}$

or $\mathbf{r} = \begin{pmatrix} 4 + 2t \\ 5 - 2t \\ -1 + 3t \end{pmatrix}$

You don't have to use λ for the parameter. In this example, the parameter is represented by the letter t.

You can give your answer in any of these forms.

Example (33)   **SKILLS**   CRITICAL THINKING

The straight line l has vector equation $\mathbf{r} = (3\mathbf{i} + 2\mathbf{j} - 5\mathbf{k}) + t(\mathbf{i} - 6\mathbf{j} - 2\mathbf{k})$
Given that the point $(a, b, 0)$ lies on l, find the value of a and the value of b.

$$\mathbf{r} = \begin{pmatrix} 3 + t \\ 2 - 6t \\ -5 - 2t \end{pmatrix}$$ You can write the equation in this form.

$-5 - 2t = 0$

$t = -\dfrac{5}{2}$

Use the z-coordinate (which is equal to zero) to find the value of t.

$a = 3 + t = \dfrac{1}{2}$

$b = 2 - 6t = 17$ Find a and b using the value of t.

$a = \dfrac{1}{2}$ and $b = 17$

Example (34)   **SKILLS**   CREATIVITY

The straight line l has vector equation $\mathbf{r} = (2\mathbf{i} + 5\mathbf{j} - 3\mathbf{k}) + \lambda(6\mathbf{i} - 2\mathbf{j} + 4\mathbf{k})$
Show that another vector equation of l is $\mathbf{r} = (8\mathbf{i} + 3\mathbf{j} + \mathbf{k}) + \mu(3\mathbf{i} - \mathbf{j} + 2\mathbf{k})$

Use the equation $\mathbf{r} = \begin{pmatrix} 2 \\ 5 \\ -3 \end{pmatrix} + \lambda \begin{pmatrix} 6 \\ -2 \\ 4 \end{pmatrix}$

When $\lambda = 1$, $\mathbf{r} = \begin{pmatrix} 8 \\ 3 \\ 1 \end{pmatrix}$, so the point $(8, 3, 1)$ lies on l.

$\begin{pmatrix} 6 \\ -2 \\ 4 \end{pmatrix} = 2 \begin{pmatrix} 3 \\ -1 \\ 2 \end{pmatrix}$ so these two vectors are parallel.

So an alternative form of the equation is

$\mathbf{r} = \begin{pmatrix} 8 \\ 3 \\ 1 \end{pmatrix} + \mu \begin{pmatrix} 3 \\ -1 \\ 2 \end{pmatrix}$

To show that $(8\mathbf{i} + 3\mathbf{j} + \mathbf{k})$ lies on l, find a value of λ that gives this point. It is often easier to work in column vectors.

If one vector is a scalar multiple of another then the vectors are parallel.

Watch out Using the same value of the parameter in each equation will give **different** points on the line. You should use a different letter for the parameter of the second equation.

Example 35 SKILLS CRITICAL THINKING

The line l has equation $\mathbf{r} = \begin{pmatrix} -2 \\ 1 \\ 4 \end{pmatrix} + \lambda \begin{pmatrix} 1 \\ -2 \\ 1 \end{pmatrix}$, and the point P has position vector $\begin{pmatrix} 2 \\ 1 \\ 3 \end{pmatrix}$

a Show that P does not lie on l.

Given that a circle, centre P, intersects l at points A and B, and that A has position vector $\begin{pmatrix} 0 \\ -3 \\ 6 \end{pmatrix}$

b find the position vector of B.

a $\mathbf{r} = \begin{pmatrix} -2 + \lambda \\ 1 - 2\lambda \\ 4 + \lambda \end{pmatrix}$

If $P(2, 1, 3)$ lies on the line then

$2 = -2 + \lambda \Rightarrow \lambda = 4$

$1 = 1 - 2\lambda \Rightarrow \lambda = 0$

$3 = 4 + \lambda \Rightarrow \lambda = -1$

so P does not lie on l.

b $\overrightarrow{AP} = \begin{pmatrix} 2 \\ 1 \\ 3 \end{pmatrix} - \begin{pmatrix} 0 \\ -3 \\ 6 \end{pmatrix} = \begin{pmatrix} 2 \\ 4 \\ -3 \end{pmatrix}$

$|\overrightarrow{AP}| = \sqrt{2^2 + 4^2 + (-3)^2} = \sqrt{29}$

Problem-solving

It is often useful to write the general point on a line as a single vector. You can write each component in the form $a + \lambda b$

If P lies on l, there is one value of λ that satisfies all 3 equations. You only need to show that two of these equations are not consistent to show that P does not lie on l.

The distance between the points with position vectors $\begin{pmatrix} a_1 \\ a_2 \\ a_3 \end{pmatrix}$ and $\begin{pmatrix} b_1 \\ b_2 \\ b_3 \end{pmatrix}$ is

$\sqrt{(b_1 - a_1)^2 + (b_2 - a_2)^2 + (b_3 - a_3)^2}$. As P is the centre of the circle and A lies on the circle, the radius of the circle is $\sqrt{29}$

← Pure Year 2, Chapter 12

The position vector of B is $\begin{pmatrix} -2 + \lambda \\ 1 - 2\lambda \\ 4 + \lambda \end{pmatrix}$

$\overrightarrow{BP} = \begin{pmatrix} 2 \\ 1 \\ 3 \end{pmatrix} - \begin{pmatrix} -2 + \lambda \\ 1 - 2\lambda \\ 4 + \lambda \end{pmatrix} = \begin{pmatrix} 4 - \lambda \\ 2\lambda \\ -1 - \lambda \end{pmatrix}$

$(4 - \lambda)^2 + 4\lambda^2 + (-1 - \lambda)^2 = 29$

$16 - 8\lambda + \lambda^2 + 4\lambda^2 + 1 + 2\lambda + \lambda^2 = 29$

Use the general point on the line to represent the position vector of B.

B lies on the circle so the length $\left| \overrightarrow{BP} \right| = \sqrt{29}$

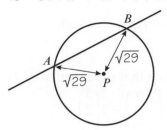

Solve the resulting quadratic equation to find two possible values of λ. One will correspond to point A, and the other will correspond to point B.

$6\lambda^2 - 6\lambda + 17 = 29$

$6\lambda^2 - 6\lambda - 12 = 0$

$\lambda^2 - \lambda - 2 = 0$

$(\lambda - 2)(\lambda + 1) = 0$

So $\lambda = 2$ or $\lambda = -1$

$\lambda = 2$ gives $\begin{pmatrix} 0 \\ -3 \\ 6 \end{pmatrix}$

This is the position vector of point A.

$\lambda = -1$ gives $\begin{pmatrix} -3 \\ 3 \\ 3 \end{pmatrix}$

This is the position vector of point B.

Substitute values of λ into $\begin{pmatrix} -2 + \lambda \\ 1 - 2\lambda \\ 4 + \lambda \end{pmatrix}$

Check that one of the values gives the position vector of A. The other value must give the position vector for B.

Exercise **7I** **SKILLS** **CRITICAL THINKING**

1 For the following pairs of vectors, find a vector equation of the straight line which passes through the point, with position vector **a**, and is parallel to the vector **b**.

a $\mathbf{a} = 6\mathbf{i} + 5\mathbf{j} - \mathbf{k}$, $\mathbf{b} = 2\mathbf{i} - 3\mathbf{j} - \mathbf{k}$

b $\mathbf{a} = 2\mathbf{i} + 5\mathbf{j}$, $\mathbf{b} = \mathbf{i} + \mathbf{j} + \mathbf{k}$

c $\mathbf{a} = -7\mathbf{i} + 6\mathbf{j} + 2\mathbf{k}$, $\mathbf{b} = 3\mathbf{i} + \mathbf{j} + 2\mathbf{k}$

d $\mathbf{a} = \begin{pmatrix} 2 \\ 0 \\ 4 \end{pmatrix}$, $\mathbf{b} = \begin{pmatrix} -3 \\ 2 \\ 1 \end{pmatrix}$

e $\mathbf{a} = \begin{pmatrix} 6 \\ -11 \\ 2 \end{pmatrix}$, $\mathbf{b} = \begin{pmatrix} 0 \\ 5 \\ -2 \end{pmatrix}$

2 For the points P and Q with position vectors $\mathbf{p}$ and $\mathbf{q}$ respectively, find

   i the vector $\overrightarrow{PQ}$

   ii a vector equation of the straight line that passes through P and Q

   a $\mathbf{p} = 3\mathbf{i} - 4\mathbf{j} + 2\mathbf{k}, \mathbf{q} = 5\mathbf{i} + 3\mathbf{j} - \mathbf{k}$ **b** $\mathbf{p} = 2\mathbf{i} + \mathbf{j} - 3\mathbf{k}, \mathbf{q} = 4\mathbf{i} - 2\mathbf{j} + \mathbf{k}$

   c $\mathbf{p} = \mathbf{i} - 2\mathbf{j} + 4\mathbf{k}, \mathbf{q} = -2\mathbf{i} - 3\mathbf{j} + 2\mathbf{k}$ **d** $\mathbf{p} = \begin{pmatrix} 3 \\ -1 \\ 4 \end{pmatrix}, \mathbf{q} = \begin{pmatrix} -2 \\ 3 \\ 1 \end{pmatrix}$

   e $\mathbf{p} = \begin{pmatrix} 4 \\ -2 \\ 3 \end{pmatrix}, \mathbf{q} = \begin{pmatrix} -2 \\ 2 \\ 4 \end{pmatrix}$

3 Find a vector equation of the line which is parallel to the z-axis and passes through the point $(4, -3, 8)$

(P) 4 The point $(1, p, q)$ lies on the line l. Find the values of p and q, given that the equation of l is

   a $\mathbf{r} = \begin{pmatrix} 2 \\ -3 \\ 1 \end{pmatrix} + \lambda\begin{pmatrix} 1 \\ -4 \\ -9 \end{pmatrix}$ **b** $\mathbf{r} = \begin{pmatrix} -4 \\ 6 \\ -1 \end{pmatrix} + \lambda\begin{pmatrix} 2 \\ -5 \\ -8 \end{pmatrix}$ **c** $\mathbf{r} = \begin{pmatrix} 16 \\ -9 \\ -10 \end{pmatrix} + \lambda\begin{pmatrix} 3 \\ 2 \\ 1 \end{pmatrix}$

(P) 5 Show that the line l_1 with equation $\mathbf{r} = (3 + 2\lambda)\mathbf{i} + (2 - 3\lambda)\mathbf{j} + (-1 + 4\lambda)\mathbf{k}$ is parallel to the line l_2 which passes through the points $A(5, 4, -1)$ and $B(3, 7, -5)$

6 Show that the points $A(-3, -4, 5)$, $B(3, -1, 2)$ and $C(9, 2, -1)$ are **collinear**.

> **Hint** Points are said to be **collinear** if they all lie on the same straight line.

7 Show that the points with position vectors $\begin{pmatrix} 1 \\ 7 \\ -2 \end{pmatrix}$, $\begin{pmatrix} 3 \\ -1 \\ 8 \end{pmatrix}$ and $\begin{pmatrix} 10 \\ 4 \\ 0 \end{pmatrix}$ do not lie on the same straight line.

(E/P) 8 The points $P(2, 0, 4)$, $Q(a, 5, 1)$ and $R(3, 10, b)$, where a and b are constants, are collinear. Find the values of a and b. **(5 marks)**

(E) 9 The line l_1 has equation

    $\mathbf{r} = (8\mathbf{i} - 5\mathbf{j} + 4\mathbf{k}) + \lambda(3\mathbf{i} + \mathbf{j} - 6\mathbf{k})$

   A is the point on l_1 such that $\lambda = -2$

   The line l_2 passes through A and is parallel to the line with equation

    $\mathbf{r} = (10\mathbf{i} + 3\mathbf{j} - 9\mathbf{k}) + \lambda(2\mathbf{i} - 4\mathbf{j} + \mathbf{k})$

   Find an equation for l_2. **(6 marks)**

(E/P) 10 The point A with coordinates $(4, a, 0)$ lies on the line L with vector equation

    $\mathbf{r} = (10\mathbf{i} + 8\mathbf{j} - 12\mathbf{k}) + \lambda(\mathbf{i} - \mathbf{j} + b\mathbf{k})$

   where a and b are constants.

   a Find the values of a and b. **(3 marks)**

   The point X lies on L where $\lambda = -1$.

   b Find the coordinates of X. **(1 mark)**

(E) **11** The line l has equation $\mathbf{r} = \begin{pmatrix} 3 \\ -5 \\ 9 \end{pmatrix} + \lambda \begin{pmatrix} 1 \\ 2 \\ -2 \end{pmatrix}$

A and B are the points on l with $\lambda = 5$ and $\lambda = 2$ respectively.
Find the distance AB.   **(4 marks)**

(E) **12** The line l has equation $\mathbf{r} = \begin{pmatrix} 1 \\ -2 \\ 3 \end{pmatrix} + \lambda \begin{pmatrix} 2 \\ 1 \\ -1 \end{pmatrix}$

C and A are the points on l with $\lambda = 4$ and $\lambda = 3$ respectively.
A circle has centre C and intersects l at the points A and B.
Find the position vector of B.   **(3 marks)**

(E/P) **13** The line l_1 has equation $\mathbf{r} = \begin{pmatrix} -4 \\ 6 \\ 5 \end{pmatrix} + \lambda \begin{pmatrix} 1 \\ -1 \\ 1 \end{pmatrix}$

A and B are the points on l_1 with $\lambda = 2$ and $\lambda = 5$ respectively.

a Find the position vectors of A and B.   **(2 marks)**

The point P has position vector $\begin{pmatrix} 0 \\ 2 \\ 3 \end{pmatrix}$

The line l_2 passes through the point P and is parallel to the line l_1.
b Find a vector equation of the line l_2.   **(2 marks)**
The points C and D both lie on line l_2 such that $AB = AC = AD$.
c Show that P is the midpoint of CD.   **(7 marks)**

(E/P) **14** A tightrope is modelled as a line segment between points with coordinates (2, 3, 8) and (22, 18, 8), relative to a fixed origin O, where the units of distance are metres. Two support cables are **anchored** to a fixed point A on the wire. The other ends of the cables are anchored to points with coordinates (14, 1, 0) and (6, 17, 0) respectively.
a Given that the support cables are both 12 m long, find the coordinates of A.   **(8 marks)**
b Give one criticism of this model.   **(1 mark)**

7.10   Points of intersection

You need to be able to determine whether two lines meet and, if so, to determine their point of intersection.

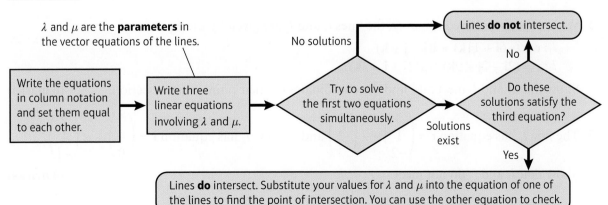

Example (36)   **SKILLS**   REASONING

The lines l_1 and l_2 have vector equations

$\mathbf{r} = 3\mathbf{i} + \mathbf{j} + \mathbf{k} + \lambda(\mathbf{i} - 2\mathbf{j} - \mathbf{k})$ and $\mathbf{r} = -2\mathbf{j} + 3\mathbf{k} + \mu(-5\mathbf{i} + \mathbf{j} + 4\mathbf{k})$ respectively.

Show that the two lines intersect, and find the position vector of the point of intersection.

$$\begin{pmatrix} 3 + \lambda \\ 1 - 2\lambda \\ 1 - \lambda \end{pmatrix} = \begin{pmatrix} -5\mu \\ -2 + \mu \\ 3 + 4\mu \end{pmatrix}$$

Use column vector notation for clarity, and to help to avoid errors.

Solve the simultaneous equations

$3 + \lambda = -5\mu$ (1)

and $1 - \lambda = 3 + 4\mu$ (2)

Choose two of the three equations obtained by equating x-, y- and z-components and solve the resulting simultaneous equations.

Adding gives $4 = 3 - \mu$

and so $\mu = -1$

Substituting back into equation (1) gives $\lambda = 2$

Check $\mu = -1$, $\lambda = 2$ also satisfy the third equation.

$1 - 2\lambda = -2 + \mu$ gives $-3 = -3$

If the lines intersect there is a pair of values of λ and μ that satisfy the 3 equations simultaneously.

So the lines do intersect.

Substituting $\lambda = 2$ into $\begin{pmatrix} 3 + \lambda \\ 1 - 2\lambda \\ 1 - \lambda \end{pmatrix}$ gives $\begin{pmatrix} 5 \\ -3 \\ -1 \end{pmatrix}$.

The point where the lines meet is $(5, -3, -1)$

Check that the point which you obtain after substitution lies on both straight lines.

Exercise (7J)

1 In each case establish whether lines l_1 and l_2 meet and, if they meet, find the coordinates of their point of intersection:

a l_1 has equation $\mathbf{r} = \mathbf{i} + 3\mathbf{j} + \lambda(\mathbf{i} - \mathbf{j} + 5\mathbf{k})$ and l_2 has equation $\mathbf{r} = -\mathbf{i} - 3\mathbf{j} + 2\mathbf{k} + \mu(\mathbf{i} + \mathbf{j} + 2\mathbf{k})$

b l_1 has equation $\mathbf{r} = 3\mathbf{i} + 2\mathbf{j} + \mathbf{k} + \lambda(\mathbf{i} + \mathbf{j} + 2\mathbf{k})$ and l_2 has equation $\mathbf{r} = 4\mathbf{i} + 3\mathbf{j} + \mu(-\mathbf{i} + \mathbf{j} - \mathbf{k})$

c l_1 has equation $\mathbf{r} = \mathbf{i} + 3\mathbf{j} + 5\mathbf{k} + \lambda(2\mathbf{i} + 3\mathbf{j} + \mathbf{k})$ and l_2 has equation $\mathbf{r} = \mathbf{i} + \frac{5}{2}\mathbf{j} + \frac{5}{2}\mathbf{k} + \mu(\mathbf{i} + \mathbf{j} - 2\mathbf{k})$

(In each of the above cases λ and μ are scalars.)

(E) **2** With respect to a fixed origin O, the lines l_1 and l_2 are given by the equations

$l_1: \mathbf{r} = (-6\mathbf{i} + 11\mathbf{k}) + \lambda(\mathbf{i} - \mathbf{j} + \mathbf{k})$
$l_2: \mathbf{r} = (2\mathbf{i} - 2\mathbf{j} + 9\mathbf{k}) + \mu(2\mathbf{i} + \mathbf{j} - 3\mathbf{k})$

Show that l_1 and l_2 meet and find the coordinates of their point of intersection. **(6 marks)**

(E) **3** The line l_1 has equation $\mathbf{r} = \begin{pmatrix} 3 \\ 1 \\ -2 \end{pmatrix} + \lambda \begin{pmatrix} 2 \\ 2 \\ 3 \end{pmatrix}$ and the line l_2 has equation $\mathbf{r} = \begin{pmatrix} 5 \\ 4 \\ 0 \end{pmatrix} + \mu \begin{pmatrix} 2 \\ 1 \\ -1 \end{pmatrix}$.

Show that l_1 and l_2 do not meet. **(4 marks)**

E/P **4** The line with vector equation $\mathbf{r} = \begin{pmatrix} 5 \\ 4 \\ -1 \end{pmatrix} + \lambda \begin{pmatrix} 3 \\ -1 \\ 2 \end{pmatrix}$ is perpendicular to the line with vector

equation $\mathbf{r} = \begin{pmatrix} 0 \\ 11 \\ 3 \end{pmatrix} + \mu \begin{pmatrix} -1 \\ p \\ p \end{pmatrix}$

    a Find the value of p. **(2 marks)**

    b Show that the two lines meet, and find the coordinates of the point of intersection. **(4 marks)**

E **5** The line l_1 has vector equation $\mathbf{r} = \begin{pmatrix} 5 \\ 2 \\ 1 \end{pmatrix} + \lambda \begin{pmatrix} -1 \\ 1 \\ 2 \end{pmatrix}$ and the line l_2 has vector equation

$\mathbf{r} = \begin{pmatrix} 4 \\ 1 \\ 1 \end{pmatrix} + \mu \begin{pmatrix} 1 \\ 0 \\ -1 \end{pmatrix}$ where λ and μ are parameters.

    Given that the lines l_1 and l_2 intersect at the point A, find the coordinates of A. **(4 marks)**

E/P **6** With respect to a fixed origin O the lines l_1 and l_2 are given by the equations

$$l_1: \mathbf{r} = \begin{pmatrix} 8 \\ 2 \\ -12 \end{pmatrix} + \lambda \begin{pmatrix} -1 \\ 3 \\ 2 \end{pmatrix} \qquad l_2: \mathbf{r} = \begin{pmatrix} -4 \\ 10 \\ p \end{pmatrix} + \mu \begin{pmatrix} q \\ 2 \\ -1 \end{pmatrix}$$

    where λ and μ are parameters and p and q are constants. Given that l_1 and l_2 are perpendicular,

    a show that $q = 4$ **(2 marks)**

    Given further that l_1 and l_2 intersect, find

    b the value of p **(6 marks)**

    c the coordinates of the point of intersection. **(2 marks)**

    The point A lies on l_1 and has position vector $\begin{pmatrix} 9 \\ -1 \\ -14 \end{pmatrix}$ The point C lies on l_2.

    Given that a circle, with centre C, cuts the line l_1 at the points A and B,

    d find the position vector of B. **(3 marks)**

> **Problem-solving**
>
> Draw a diagram showing the lines l_1 and l_2 and the circle, and use circle properties.

7.11 Scalar product

You need to know the definition of the scalar product of two vectors in either two or three dimensions, and how it can be used to find the angle between two vectors. To define the scalar product you need to know how to find the **angle between two vectors**.

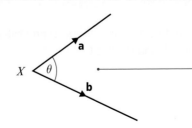

On the diagram, the angle between the vectors **a** and **b** is θ.

Notice that **a** and **b** are both directed away from the point X.

Example **37** **SKILLS** INTERPRETATION

Find the angle between the vectors **a** and **b** on the diagram.

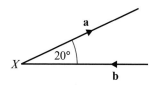

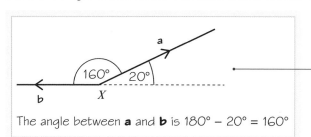

> For the correct angle, **a** and **b** must both be pointing away from X, so re-draw to show this.

The angle between **a** and **b** is $180° - 20° = 160°$

■ The scalar product of two vectors **a** and **b** is written as **a.b**, and defined as

$$\mathbf{a.b} = |\mathbf{a}||\mathbf{b}| \cos\theta$$

where θ is the angle between **a** and **b**.

> **Notation** The scalar product is often called the **dot product**. You say 'a dot b'.

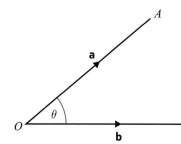

> You can see from this diagram that if **a** and **b** are the position vectors of A and B, then the angle between **a** and **b** is $\angle AOB$.

> **Online** Use GeoGebra to consider the scalar product as the component of one vector in the direction of another.

■ If **a** and **b** are the position vectors of the points A and B, then $\cos(\angle AOB) = \dfrac{\mathbf{a.b}}{|\mathbf{a}||\mathbf{b}|}$

If two vectors **a** and **b** are perpendicular, the angle between them is $90°$

Since $\cos 90° = 0$, $\mathbf{a.b} = |\mathbf{a}||\mathbf{b}| \cos 90° = 0$

■ The non-zero vectors **a** and **b** are perpendicular if and only if $\mathbf{a.b} = 0$

If **a** and **b** are parallel, the angle between them is $0°$.

■ If **a** and **b** are parallel, $\mathbf{a.b} = |\mathbf{a}||\mathbf{b}|$. In particular, $\mathbf{a.a} = |\mathbf{a}|^2$

Example **38** **SKILLS** PROBLEM-SOLVING

Find the values of

a **i.j**   **b** **k.k**   **c** $(4\mathbf{j}).\mathbf{k} + (3\mathbf{i}).(3\mathbf{i})$

a $\mathbf{i.j} = 1 \times 1 \times \cos 90° = 0$

> **i** and **j** are unit vectors (magnitude 1), and are perpendicular.

b $\mathbf{k.k} = 1 \times 1 \times \cos 0° = 1$

> **k** is a unit vector (magnitude 1) and the angle between **k** and itself is $0°$.

c $(4\mathbf{j}).\mathbf{k} + (3\mathbf{i}).(3\mathbf{i})$
$= (4 \times 1 \times \cos 90°) + (3 \times 3 \times \cos 0°)$
$= 0 + 9 = 9$

Example 39   SKILLS   REASONING

Given that $\mathbf{a} = \begin{pmatrix} a_1 \\ a_2 \\ a_3 \end{pmatrix}$ and $\mathbf{b} = \begin{pmatrix} b_1 \\ b_2 \\ b_3 \end{pmatrix}$, prove that $\mathbf{a.b} = a_1 b_1 + a_2 b_2 + a_3 b_3$

$\mathbf{a.b} = (a_1\mathbf{i} + a_2\mathbf{j} + a_3\mathbf{k}).(b_1\mathbf{i} + b_2\mathbf{j} + b_3\mathbf{k})$

$= a_1\mathbf{i}.(b_1\mathbf{i} + b_2\mathbf{j} + b_3\mathbf{k})$
$\quad + a_2\mathbf{j}.(b_1\mathbf{i} + b_2\mathbf{j} + b_3\mathbf{k})$
$\quad + a_3\mathbf{k}.(b_1\mathbf{i} + b_2\mathbf{j} + b_3\mathbf{k})$

$= (a_1\mathbf{i}).(b_1\mathbf{i}) + (a_1\mathbf{i}).(b_2\mathbf{j}) + (a_1\mathbf{i}).(b_3\mathbf{k})$
$\quad + (a_2\mathbf{j}).(b_1\mathbf{i}) + (a_2\mathbf{j}).(b_2\mathbf{j}) + (a_2\mathbf{j}).(b_3\mathbf{k})$
$\quad + (a_3\mathbf{k}).(b_1\mathbf{i}) + (a_3\mathbf{k}).(b_2\mathbf{j}) + (a_3\mathbf{k}).(b_3\mathbf{k})$

$= (a_1 b_1)\mathbf{i.i} + (a_1 b_2)\mathbf{i.j} + (a_1 b_3)\mathbf{i.k}$
$\quad + (a_2 b_1)\mathbf{j.i} + (a_2 b_2)\mathbf{j.j} + (a_2 b_3)\mathbf{j.k}$
$\quad + (a_3 b_1)\mathbf{k.i} + (a_3 b_2)\mathbf{k.j} + (a_3 b_3)\mathbf{k.k}$

$= a_1 b_1 + a_2 b_2 + a_3 b_3$

Use the results for parallel and perpendicular unit vectors:

$\mathbf{i.i} = \mathbf{j.j} = \mathbf{k.k} = 1$

$\mathbf{i.j} = \mathbf{i.k} = \mathbf{j.i} = \mathbf{j.k} = \mathbf{k.i} = \mathbf{k.j} = 0$

The above example leads to a simple formula for finding the scalar product of two vectors given in Cartesian component form:

■ If $\mathbf{a} = a_1\mathbf{i} + a_2\mathbf{j} + a_3\mathbf{k}$ and $\mathbf{b} = b_1\mathbf{i} + b_2\mathbf{j} + b_3\mathbf{k}$,

$$\mathbf{a.b} = \begin{pmatrix} a_1 \\ a_2 \\ a_3 \end{pmatrix}.\begin{pmatrix} b_1 \\ b_2 \\ b_3 \end{pmatrix} = a_1 b_1 + a_2 b_2 + a_3 b_3$$

You can use this result without proof in your exam.

Example 40   SKILLS   PROBLEM-SOLVING

Given that $\mathbf{a} = 8\mathbf{i} - 5\mathbf{j} - 4\mathbf{k}$ and $\mathbf{b} = 5\mathbf{i} + 4\mathbf{j} - \mathbf{k}$

a find $\mathbf{a.b}$

b find the angle between $\mathbf{a}$ and $\mathbf{b}$, giving your answer in degrees to 1 decimal place.

a $\mathbf{a.b} = \begin{pmatrix} 8 \\ -5 \\ -4 \end{pmatrix}.\begin{pmatrix} 5 \\ 4 \\ -1 \end{pmatrix}$

Write in column vector form.

$= (8 \times 5) + (-5 \times 4) + (-4 \times -1)$

Use $\mathbf{a.b} = a_1 b_1 + a_2 b_2 + a_3 b_3$

$= 40 - 20 + 4$

$= 24$

b $\mathbf{a}.\mathbf{b} = |\mathbf{a}||\mathbf{b}| \cos\theta$ ←——————————— Use the scalar product definition.

$|\mathbf{a}| = \sqrt{8^2 + (-5)^2 + (-4)^2} = \sqrt{105}$

$|\mathbf{b}| = \sqrt{5^2 + 4^2 + (-1)^2} = \sqrt{42}$ ←——— Find the modulus of **a** and of **b**.

$\sqrt{105}\sqrt{42} \cos\theta = 24$ ←———————————— Use $\mathbf{a}.\mathbf{b} = |\mathbf{a}||\mathbf{b}| \cos\theta$

$\cos\theta = \dfrac{24}{\sqrt{105}\sqrt{42}}$

$\theta = 68.8°$ (1 d.p.)

Example 41   **SKILLS**   PROBLEM-SOLVING

Given that $\mathbf{a} = -\mathbf{i} + \mathbf{j} + 3\mathbf{k}$ and $\mathbf{b} = 7\mathbf{i} - 2\mathbf{j} + 2\mathbf{k}$, find the angle between **a** and **b**, giving your answer in degrees to 1 decimal place.

$\mathbf{a}.\mathbf{b} = \begin{pmatrix} -1 \\ 1 \\ 3 \end{pmatrix}.\begin{pmatrix} 7 \\ -2 \\ 2 \end{pmatrix} = -7 - 2 + 6 = -3$

$|\mathbf{a}| = \sqrt{(-1)^2 + 1^2 + 3^2} = \sqrt{11}$ — For the scalar product formula, you need to find **a.b**, |**a**| and |**b**|

$|\mathbf{b}| = \sqrt{7^2 + (-2)^2 + 2^2} = \sqrt{57}$

$\sqrt{11}\sqrt{57} \cos\theta = -3$ ←———————————— Use $\mathbf{a}.\mathbf{b} = |\mathbf{a}||\mathbf{b}| \cos\theta$

$\cos\theta = \dfrac{-3}{\sqrt{11}\sqrt{57}}$ ←—————— The cosine is negative, so the angle is obtuse.

$\theta = 96.9°$ (1 d.p.)

Example 42   **SKILLS**   CRITICAL THINKING

Given that the vectors $\mathbf{a} = 2\mathbf{i} - 6\mathbf{j} + \mathbf{k}$ and $\mathbf{b} = 5\mathbf{i} + 2\mathbf{j} + \lambda\mathbf{k}$ are perpendicular, find the value of λ.

$\mathbf{a}.\mathbf{b} = \begin{pmatrix} 2 \\ -6 \\ 1 \end{pmatrix}.\begin{pmatrix} 5 \\ 2 \\ \lambda \end{pmatrix}$

$= 10 - 12 + \lambda$ ←——————————————— Find the scalar product.

$= -2 + \lambda$

$-2 + \lambda = 0$ ←—————————————— For perpendicular vectors, the scalar product is zero.

$\lambda = 2$

Example 43   **SKILLS**   INNOVATION

Given that $\mathbf{a} = -2\mathbf{i} + 5\mathbf{j} - 4\mathbf{k}$ and $\mathbf{b} = 4\mathbf{i} - 8\mathbf{j} + 5\mathbf{k}$, find a vector which is perpendicular to both $\mathbf{a}$ and $\mathbf{b}$.

$\mathbf{a} . \begin{pmatrix} x \\ y \\ z \end{pmatrix} = 0$ and $\mathbf{b} . \begin{pmatrix} x \\ y \\ z \end{pmatrix} = 0$ **Both scalar products are zero.**

$\begin{pmatrix} -2 \\ 5 \\ -4 \end{pmatrix} . \begin{pmatrix} x \\ y \\ z \end{pmatrix} = 0$ and $\begin{pmatrix} 4 \\ -8 \\ 5 \end{pmatrix} . \begin{pmatrix} x \\ y \\ z \end{pmatrix} = 0$

$\qquad -2x + 5y - 4z = 0 \qquad\qquad (1)$

$\qquad 4x - 8y + 5z = 0 \qquad\qquad (2)$

Let $z = 1$ **Choose a (non-zero) value for z (or for x, or for y).**

$\qquad -2x + 5y = 4 \qquad$ (from 1)

$\qquad 4x - 8y = -5 \qquad$ (from 2)   **Solving simutaneously gives** $x = \dfrac{7}{4}$ and $y = \dfrac{3}{2}$

So $x = \dfrac{7}{4}$, $y = \dfrac{3}{2}$ and $z = 1$

A possible vector is $\dfrac{7}{4}\mathbf{i} + \dfrac{3}{2}\mathbf{j} + \mathbf{k}$

Another possible vector is

$4\left(\dfrac{7}{4}\mathbf{i} + \dfrac{3}{2}\mathbf{j} + \mathbf{k}\right) = 7\mathbf{i} + 6\mathbf{j} + 4\mathbf{k}$ **You can multiply by a scalar constant to find another vector which is also perpendicular to both $\mathbf{a}$ and $\mathbf{b}$.**

Example 44   **SKILLS**   CRITICAL THINKING

The points A, B and C have coordinates $(2, -1, 1)$, $(5, 1, 7)$ and $(6, -3, 1)$ respectively.

a Find $\overrightarrow{AB}.\overrightarrow{AC}$

b Hence, or otherwise, find the area of triangle ABC.

a $\overrightarrow{AB} = \begin{pmatrix} 3 \\ 2 \\ 6 \end{pmatrix}$ and $\overrightarrow{AC} = \begin{pmatrix} 4 \\ -2 \\ 0 \end{pmatrix}$

$\overrightarrow{AB}.\overrightarrow{AC} = 3 \times 4 + 2 \times (-2) + 6 \times 0 = 8$

b $\left|\overrightarrow{AB}\right| = \sqrt{3^2 + 2^2 + 6^2} = 7$

$\left|\overrightarrow{AC}\right| = \sqrt{4^2 + (-2)^2 + 0^2} = 2\sqrt{5}$

$\cos(\angle BAC) = \dfrac{\overrightarrow{AB}.\overrightarrow{AC}}{\left|\overrightarrow{AB}\right|\left|\overrightarrow{AC}\right|}$ **Use the scalar product to find the angle between $\overrightarrow{AB}$ and $\overrightarrow{AC}$. Then use area $= \dfrac{1}{2}ab\sin\theta$ to find the area of the triangle.**

$\qquad\qquad = \dfrac{8}{7 \times 2\sqrt{5}}$

$\qquad\qquad = 0.2555...$

$\angle BAC = 75.1937...°$

$\text{Area} = \dfrac{1}{2}\left|\overrightarrow{AB}\right|\left|\overrightarrow{AC}\right|\sin(\angle BAC)$

$\qquad = \dfrac{1}{2} \times 7 \times 2\sqrt{5}\,\sin(75.1937...°)$

$\qquad = 15.13\ (2\ d.p.)$

Problem-solving

You could find $\angle BAC$ by finding the lengths AB, BC and AC and using the cosine rule, but it is quicker to use a vector method.

Exercise 7K   **SKILLS**   **ANALYSIS**

1 The vectors **a** and **b** each have magnitude 3, and the angle between **a** and **b** is 60°. Find **a.b**.

2 For each pair of vectors, find **a.b**.
   a $\mathbf{a} = 5\mathbf{i} + 2\mathbf{j} + 3\mathbf{k}, \mathbf{b} = 2\mathbf{i} - \mathbf{j} - 2\mathbf{k}$ **b** $\mathbf{a} = 10\mathbf{i} - 7\mathbf{j} + 4\mathbf{k}, \mathbf{b} = 3\mathbf{i} - 5\mathbf{j} - 12\mathbf{k}$
   c $\mathbf{a} = \mathbf{i} + \mathbf{j} - \mathbf{k}, \mathbf{b} = -\mathbf{i} - \mathbf{j} + 4\mathbf{k}$ **d** $\mathbf{a} = 2\mathbf{i} - \mathbf{k}, \mathbf{b} = 6\mathbf{i} - 5\mathbf{j} - 8\mathbf{k}$
   e $\mathbf{a} = 3\mathbf{j} + 9\mathbf{k}, \mathbf{b} = \mathbf{i} - 12\mathbf{j} + 4\mathbf{k}$

3 In each part, find the angle between **a** and **b**, giving your answer in degrees to 1 decimal place.
   a $\mathbf{a} = 3\mathbf{i} + 7\mathbf{j}, \mathbf{b} = 5\mathbf{i} + \mathbf{j}$ **b** $\mathbf{a} = 2\mathbf{i} - 5\mathbf{j}, \mathbf{b} = 6\mathbf{i} + 3\mathbf{j}$
   c $\mathbf{a} = \mathbf{i} - 7\mathbf{j} + 8\mathbf{k}, \mathbf{b} = 12\mathbf{i} + 2\mathbf{j} + \mathbf{k}$ **d** $\mathbf{a} = -\mathbf{i} - \mathbf{j} + 5\mathbf{k}, \mathbf{b} = 11\mathbf{i} - 3\mathbf{j} + 4\mathbf{k}$
   e $\mathbf{a} = 6\mathbf{i} - 7\mathbf{j} + 12\mathbf{k}, \mathbf{b} = -2\mathbf{i} + \mathbf{j} + \mathbf{k}$ **f** $\mathbf{a} = 4\mathbf{i} + 5\mathbf{j}, \mathbf{b} = 6\mathbf{i} - 2\mathbf{j}$
   g $\mathbf{a} = -5\mathbf{i} + 2\mathbf{j} - 3\mathbf{k}, \mathbf{b} = 2\mathbf{i} - 2\mathbf{j} - 11\mathbf{k}$ **h** $\mathbf{a} = \mathbf{i} + \mathbf{j} + \mathbf{k}, \mathbf{b} = \mathbf{i} - \mathbf{j} + \mathbf{k}$

4 Find the value, or values, of λ for which these vectors are perpendicular.
   a $3\mathbf{i} + 5\mathbf{j}$ and $\lambda\mathbf{i} + 6\mathbf{j}$ **b** $2\mathbf{i} + 6\mathbf{j} - \mathbf{k}$ and $\lambda\mathbf{i} - 4\mathbf{j} - 14\mathbf{k}$
   c $3\mathbf{i} + \lambda\mathbf{j} - 8\mathbf{k}$ and $7\mathbf{i} - 5\mathbf{j} + \mathbf{k}$ **d** $9\mathbf{i} - 3\mathbf{j} + 5\mathbf{k}$ and $\lambda\mathbf{i} + \lambda\mathbf{j} + 3\mathbf{k}$
   e $\lambda\mathbf{j} + 3\mathbf{j} - 2\mathbf{k}$ and $\lambda\mathbf{i} + \lambda\mathbf{j} + 5\mathbf{k}$

5 Find, to the nearest tenth of a degree, the angle that the vector $9\mathbf{i} - 5\mathbf{j} + 3\mathbf{k}$ makes with
   a the positive x-axis   **b** the positive y-axis

6 Find, to the nearest tenth of a degree, the angle that the vector $\mathbf{i} + 11\mathbf{j} - 4\mathbf{k}$ makes with
   a the positive y-axis   **b** the positive z-axis

7 The angle between the vectors $\mathbf{i} + \mathbf{j} + \mathbf{k}$ and $2\mathbf{i} + \mathbf{j} + \mathbf{k}$ is θ. Calculate the exact value of $\cos\theta$.

8 The angle between the vectors $\mathbf{i} + 3\mathbf{j}$ and $\mathbf{j} + \lambda\mathbf{k}$ is 60°. Show that $\lambda = \pm\sqrt{\dfrac{13}{5}}$

9 Find a vector which is perpendicular to both **a** and **b**, where
   a $\mathbf{a} = \mathbf{i} + \mathbf{j} - 3\mathbf{k}, \mathbf{b} = 5\mathbf{i} - 2\mathbf{j} - \mathbf{k}$ **b** $\mathbf{a} = 2\mathbf{i} + 3\mathbf{j} - 4\mathbf{k}, \mathbf{b} = \mathbf{i} - 6\mathbf{j} + 3\mathbf{k}$
   c $\mathbf{a} = 4\mathbf{i} - 4\mathbf{j} - \mathbf{k}, \mathbf{b} = -2\mathbf{i} - 9\mathbf{j} + 6\mathbf{k}$

10 The points A and B have position vectors $2\mathbf{i} + 5\mathbf{j} + \mathbf{k}$ and $6\mathbf{i} + \mathbf{j} - 2\mathbf{k}$ respectively, and O is the origin. Calculate each of the angles in $\triangle OAB$, giving your answers in degrees to 1 decimal place.

11 The points A, B and C have coordinates $(1, 3, 1)$, $(2, 7, -3)$ and $(4, -5, 2)$ respectively.

    a Find the exact lengths of AB and BC.

    b Calculate, to one decimal place, the size of $\angle ABC$.

(E/P) 12 Given that the points A and B have coordinates $(7, 4, 4)$ and $(2, 2, 1)$ respectively,

    a find the value of $\cos \angle AOB$, where O is the origin   **(4 marks)**

    b show that the area of $\triangle AOB$ is $\dfrac{\sqrt{53}}{2}$ **(3 marks)**

(P) 13 AB is a diameter of a circle centred at the origin O, and P is a point on the circumference of the circle. By considering the position vectors of A, B and P, prove that AP is perpendicular to BP.

> **Problem-solving**
>
> This is a vector proof of the fact that the angle in a semi-circle is 90°.

(E/P) 14 Points A, B and C have coordinates $(5, -1, 0)$, $(2, 4, 10)$ and $(6, -1, 4)$ respectively.

    a Find the vectors $\overrightarrow{CA}$ and $\overrightarrow{CB}$ **(2 marks)**

    b Find the area of the triangle ABC.   **(4 marks)**

    c Point D is such that A, B, C and D are the vertices of a parallelogram. Find the coordinates of three possible positions of D.   **(3 marks)**

    d Write down the area of the parallelogram.   **(1 mark)**

(E/P) 15 The points P, Q and R have coordinates $(1, -1, 6)$, $(-2, 5, 4)$ and $(0, 3, -5)$ respectively.

    a Show that PQ is perpendicular to QR.   **(3 marks)**

    b Hence find the centre and radius of the circle that passes through points P, Q and R.   **(3 marks)**

Challenge

1 Using the definition $\mathbf{a}.\mathbf{b} = |\mathbf{a}||\mathbf{b}| \cos \theta$, prove that $\mathbf{a}.\mathbf{b} = \mathbf{b}.\mathbf{a}$

2 The diagram shows arbitrary vectors $\mathbf{a}$, $\mathbf{b}$ and $\mathbf{c}$, and the vector $\mathbf{b} + \mathbf{c}$

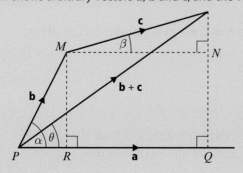

    a Show that

       i $\mathbf{a}.(\mathbf{b} + \mathbf{c}) = |\mathbf{a}| \times PQ$

       ii $\mathbf{a}.\mathbf{b} = |\mathbf{a}| \times PR$

       iii $\mathbf{a}.\mathbf{c} = |\mathbf{a}| \times RQ$

    b Hence prove that $\mathbf{a}.(\mathbf{b} + \mathbf{c}) = \mathbf{a}.\mathbf{b} + \mathbf{a}.\mathbf{c}$

Chapter review 7   **SKILLS**   PROBLEM-SOLVING

(E) **1** Two forces $\mathbf{F_1}$ and $\mathbf{F_2}$ act on a particle.

$\mathbf{F_1} = -3\mathbf{i} + 7\mathbf{j}$ newtons

$\mathbf{F_2} = \mathbf{i} - \mathbf{j}$ newtons

The resultant force $\mathbf{R}$ acting on the particle is given by $\mathbf{R} = \mathbf{F_1} + \mathbf{F_2}$

  a Calculate the magnitude of $\mathbf{R}$ in newtons.   **(3 marks)**

  b Calculate, to the nearest degree, the angle between the line of action of $\mathbf{R}$ and the vector $\mathbf{j}$.   **(2 marks)**

(P) **2** A small boat S, drifting in the sea, is modelled as a particle moving in a straight line at constant speed. When first sighted at 09:00, S is at a point with position vector $(-2\mathbf{i} - 4\mathbf{j})$ km relative to a fixed origin O, where $\mathbf{i}$ and $\mathbf{j}$ are unit vectors due east and due north respectively. At 09:40, S is at the point with position vector $(4\mathbf{i} - 6\mathbf{j})$ km.

  a Calculate the **bearing** on which S is drifting.

  b Find the speed of S.

(P) **3** $ABCD$ is a trapezium with AB parallel to DC and $DC = 4AB$

M divides DC such that $DM:MC = 3:2$, $\overrightarrow{AB} = \mathbf{a}$ and $\overrightarrow{BC} = \mathbf{b}$

Find, in terms of $\mathbf{a}$ and $\mathbf{b}$

  a $\overrightarrow{AM}$ **b** $\overrightarrow{BD}$ **c** $\overrightarrow{MB}$ **d** $\overrightarrow{DA}$

(E/P) **4** The vectors $5\mathbf{a} + k\mathbf{b}$ and $8\mathbf{a} + 2\mathbf{b}$ are parallel. Find the value of k.   **(3 marks)**

5 Given that $\mathbf{a} = \begin{pmatrix} 7 \\ 4 \end{pmatrix}$, $\mathbf{b} = \begin{pmatrix} 10 \\ -2 \end{pmatrix}$ and $\mathbf{c} = \begin{pmatrix} -5 \\ -3 \end{pmatrix}$ find

  a $\mathbf{a} + \mathbf{b} + \mathbf{c}$ **b** $\mathbf{a} - 2\mathbf{b} + \mathbf{c}$ **c** $2\mathbf{a} + 2\mathbf{b} - 3\mathbf{c}$

(E/P) **6** The resultant of the vectors $\mathbf{a} = 4\mathbf{i} - 3\mathbf{j}$ and $\mathbf{b} = 2p\mathbf{i} - p\mathbf{j}$ is parallel to the vector $\mathbf{c} = 2\mathbf{i} - 3\mathbf{j}$. Find

  a the value of p **(3 marks)**

  b the resultant of vectors $\mathbf{a}$ and $\mathbf{b}$.   **(1 mark)**

(P) **7** The vector $\mathbf{a} = p\mathbf{i} + q\mathbf{j}$, where p and q are positive constants, is such that $|\mathbf{a}| = 15$ Given that $\mathbf{a}$ makes an angle of $55°$ with $\mathbf{i}$, find the values of p and q.

(E/P) **8** Given that $|3\mathbf{i} - k\mathbf{j}| = 3\sqrt{5}$, find the value of k.   **(3 marks)**

(E/P) **9** OAB is a triangle. $\overrightarrow{OA} = \mathbf{a}$ and $\overrightarrow{OB} = \mathbf{b}$. The point M divides OA in the ratio $3:2$
MN is parallel to OB.

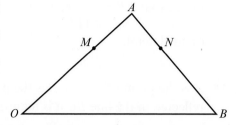

    a Express the vector $\overrightarrow{ON}$ in terms of $\mathbf{a}$ and $\mathbf{b}$.   **(4 marks)**

    b Find vector $\overrightarrow{MN}$ **(2 marks)**

    c Show that $AN:NB = 2:3$ **(2 marks)**

(P) **10** The points $A(2, 7, 3)$ and $B(4, 3, 5)$ are joined to form the line segment AB. The point M is the midpoint of AB. Find the distance from M to the point $C(5, 8, 7)$

(P) **11** The coordinates of P and Q are $(2, 3, a)$ and $(a - 2, 6, 7)$
Given that the distance from P to Q is $\sqrt{14}$, find the possible values of a.

(P) **12** $\overrightarrow{AB}$ is the vector $-3\mathbf{i} + t\mathbf{j} + 5\mathbf{k}$, where $t > 0$
Given that $|\overrightarrow{AB}| = 5\sqrt{2}$, show that $\overrightarrow{AB}$ is parallel to $6\mathbf{i} - 8\mathbf{j} - \dfrac{5}{2}t\mathbf{k}$

(P) **13** P is the point $(5, 6, -2)$, Q is the point $(2, -2, 1)$ and R is the point $(2, -3, 6)$

    a Find the vectors $\overrightarrow{PQ}$, $\overrightarrow{PR}$ and $\overrightarrow{QR}$.

    b Hence, or otherwise, find the area of triangle PQR.

(E/P) **14** The points D, E and F have position vectors $\begin{pmatrix} 1 \\ 0 \\ 0 \end{pmatrix}$, $\begin{pmatrix} 5 \\ 3 \\ 4 \end{pmatrix}$ and $\begin{pmatrix} 2 \\ -1 \\ 8 \end{pmatrix}$ respectively.

    a Find the vectors $\overrightarrow{DE}$, $\overrightarrow{EF}$ and $\overrightarrow{FD}$ **(3 marks)**

    b Find $|\overrightarrow{DE}|$, $|\overrightarrow{EF}|$ and $|\overrightarrow{FD}|$ giving your answers in exact form.   **(6 marks)**

    c Describe triangle DEF.   **(1 mark)**

(E) **15** P is the point $(-6, 2, 1)$, Q is the point $(3, -2, 1)$ and R is the point $(1, 3, -2)$

    a Find the vectors $\overrightarrow{PQ}$, $\overrightarrow{PR}$ and $\overrightarrow{QR}$ **(3 marks)**

    b Hence find the lengths of the sides of triangle PQR.   **(6 marks)**

    c Given that angle $QRP = 90°$ find the size of angle PQR.   **(2 marks)**

(E/P) **16** The diagram shows the triangle ABC.

Given that $\overrightarrow{AB} = -\mathbf{i} + \mathbf{j}$ and $\overrightarrow{BC} = \mathbf{i} - 3\mathbf{j} + \mathbf{k}$
find $\angle ABC$ to 1 d.p.

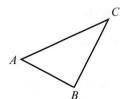

                                                                   (5 marks)

(E/P) **17** The diagram shows the quadrilateral $ABCD$.

Given that $\overrightarrow{AB} = \begin{pmatrix} 6 \\ -2 \\ 11 \end{pmatrix}$ and $\overrightarrow{AC} = \begin{pmatrix} 15 \\ 8 \\ 5 \end{pmatrix}$, find the area of the

quadrilateral.   **(7 marks)**

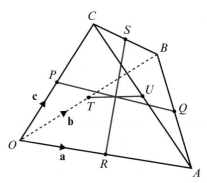

(P) **18** A is the point $(2, 3, -2)$, B is the point $(0, -2, 1)$ and C is the point $(4, -2, -5)$. When A is reflected in the line BC it is mapped to the point D.

  a Work out the coordinates of the point D.

  b Give the mathematical name for the shape $ABCD$.

  c Work out the area of $ABCD$.

(P) **19** The diagram shows a tetrahedron $OABC$. $\mathbf{a}$, $\mathbf{b}$ and $\mathbf{c}$ are the position vectors of A, B and C respectively.

P, Q, R, S, T and U are the midpoints of OC, AB, OA, BC, OB and AC respectively.

Prove that the line segments PQ, RS and TU meet at a point and bisect each other.

(P) **20** A particle of mass 2 kg is acted upon by three forces:

$\quad \mathbf{F_1} = (b\mathbf{i} + 2\mathbf{j} + \mathbf{k})\,\mathrm{N}$
$\quad \mathbf{F_2} = (3\mathbf{i} - b\mathbf{j} + 2\mathbf{k})\,\mathrm{N}$
$\quad \mathbf{F_3} = (-2\mathbf{i} + 2\mathbf{j} + (4 - b)\mathbf{k})\,\mathrm{N}$

Given that the particle accelerates at $3.5\,\mathrm{m\,s^{-2}}$, work out the possible values of b.

(P) **21** In this question $\mathbf{i}$ and $\mathbf{j}$ are the unit vectors due east and due north respectively, and $\mathbf{k}$ is the unit vector acting vertically upwards.

A BASE jumper descending with a parachute is modelled as a particle of mass 50 kg subject to forces describing the wind, $\mathbf{W}$, and air resistance, $\mathbf{F}$, where:

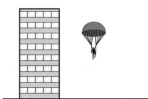

$\quad \mathbf{W} = (20\mathbf{i} + 16\mathbf{j})\,\mathrm{N}$
$\quad \mathbf{F} = (-4\mathbf{i} - 3\mathbf{j} + 450\mathbf{k})\,\mathrm{N}$

  a With reference to the model, suggest a reason why the $\mathbf{k}$ component of $\mathbf{F}$ is greater than the other components.

  b Taking $g = 9.8\,\mathrm{m\,s^{-2}}$, find the resultant force acting on the BASE jumper.

  c Given that the BASE jumper starts from rest and travels a distance of 180 m before landing, find the total time of the descent.

(E) 22 The line l passes through the points A and B with position vectors $\mathbf{i} - \mathbf{j} + 3\mathbf{k}$ and $\mathbf{i} + 2\mathbf{j} + 2\mathbf{k}$ respectively, relative to a fixed origin O.

    a Find a vector equation of the line l. **(4 marks)**

    b Find the position vector of the point C which lies on the line segment AB such that
$AC = 2CB$ **(3 marks)**

(E) 23 Find a vector equation of the straight line which passes through the point A with position vector $2\mathbf{i} + 3\mathbf{j} - 4\mathbf{k}$, and is parallel to the vector $2\mathbf{j} + 3\mathbf{k}$ **(3 marks)**

24 A straight line l has vector equation $\mathbf{r} = (\mathbf{i} + 2\mathbf{j} - \mathbf{k}) + \lambda(3\mathbf{i} + \mathbf{j} - 2\mathbf{k})$
Show that another vector equation of l is $\mathbf{r} = (7\mathbf{i} + 4\mathbf{j} - 5\mathbf{k}) + \lambda(9\mathbf{i} + 3\mathbf{j} - 6\mathbf{k})$

(E/P) 25 With respect to an origin O, the position vectors of the points L, M and N are $\begin{pmatrix} 4 \\ 7 \\ 7 \end{pmatrix}$, $\begin{pmatrix} 1 \\ 3 \\ 2 \end{pmatrix}$ and $\begin{pmatrix} 2 \\ 4 \\ 6 \end{pmatrix}$ respectively.

    a Find the vectors $\overrightarrow{ML}$ and $\overrightarrow{MN}$ **(3 marks)**

    b Prove that $\cos \angle LMN = \dfrac{9}{10}$ **(3 marks)**

(E/P) 26 Referred to a fixed origin O, the points A, B and C have position vectors $9\mathbf{i} - 2\mathbf{j} + \mathbf{k}$ $6\mathbf{i} + 2\mathbf{j} + 6\mathbf{k}$ and $3\mathbf{i} + p\mathbf{j} + q\mathbf{k}$ respectively, where p and q are constants.

    a Find, in vector form, an equation of the line l which passes through A and B. **(2 marks)**

    Given that C lies on l,

    b find the value of p and the value of q **(2 marks)**

    c calculate, in degrees, the acute angle between OC and AB. **(3 marks)**

    The point D lies on AB and is such that OD is perpendicular to AB.

    d Find the position vector of D. **(5 marks)**

(E) 27 Referred to a fixed origin O, the points A and B have position vectors $\begin{pmatrix} 1 \\ 2 \\ -3 \end{pmatrix}$ and $\begin{pmatrix} 5 \\ 0 \\ -3 \end{pmatrix}$ respectively.

    a Find, in vector form, an equation of the line l_1 which passes through A and B. **(3 marks)**

    The line l_2 has equation $\mathbf{r} = \begin{pmatrix} 4 \\ -4 \\ 3 \end{pmatrix} + \mu\begin{pmatrix} 1 \\ -2 \\ 2 \end{pmatrix}$, where μ is a scalar parameter.

    b Show that A lies on l_2. **(2 marks)**

    c Find, in degrees, the acute angle between the lines l_1 and l_2. **(4 marks)**

    The point C with position vector $\begin{pmatrix} 0 \\ 4 \\ -5 \end{pmatrix}$ lies on l_2.

    d Find the shortest distance from C to the line l_1. **(4 marks)**

(E) **28** Two submarines are travelling in straight lines through the ocean. Relative to a fixed origin, the vector equations of the two lines, l_1 and l_2, along which they travel are

$l_1: \mathbf{r} = 3\mathbf{i} + 4\mathbf{j} - 5\mathbf{k} + \lambda(\mathbf{i} - 2\mathbf{j} + 2\mathbf{k})$
$l_2: \mathbf{r} = 9\mathbf{i} + \mathbf{j} - 2\mathbf{k} + \mu(4\mathbf{i} + \mathbf{j} - \mathbf{k})$

where λ and μ are scalars.

a Show that the submarines are moving in perpendicular directions. **(2 marks)**

b Given that l_1 and l_2 intersect at the point A, find the position vector of A. **(4 marks)**

The point B has position vector $10\mathbf{j} - 11\mathbf{k}$

c Show that only one of the submarines passes through the point B. **(3 marks)**

d Given that 1 unit on each coordinate axis represents 100 m, find, in km, the distance AB. **(2 marks)**

(E/P) **29** With respect to a fixed origin O, the straight lines l_1 and l_2 are given by

$$l_1: \mathbf{r} = \begin{pmatrix} 1 \\ 1 \\ 0 \end{pmatrix} + \lambda \begin{pmatrix} 2 \\ 1 \\ -2 \end{pmatrix}$$

$$l_2: \mathbf{r} = \begin{pmatrix} 1 \\ 4 \\ -4 \end{pmatrix} + \mu \begin{pmatrix} -3 \\ 0 \\ 1 \end{pmatrix}$$

where λ and μ are scalar parameters.

a Show that the lines intersect. **(3 marks)**

b Find the position vector of their point of intersection. **(1 mark)**

c Find the cosine of the acute angle between the lines. **(4 marks)**

(E/P) **30** The line l_1 has vector equation $\mathbf{r} = 6\mathbf{i} + 8\mathbf{j} + 5\mathbf{k} + \lambda(\mathbf{i} - \mathbf{j} + \mathbf{k})$ where λ is a scalar parameter. The point A has coordinates $(3, a, 2)$, where a is a constant. The point B has coordinates $(8, 6, b)$, where b is a constant. Points A and B lie on the line l_1.

a Find the values of a and b. **(3 marks)**

Given that the point O is the origin, and that the point P lies on l_1 such that OP is perpendicular to l_1,

b find the coordinates of P. **(5 marks)**

c Hence find the distance OP, giving your answer in surd form. **(2 marks)**

(E/P) **31** Relative to a fixed origin O, the point A has position vector $6\mathbf{i} + 3\mathbf{j} + 4\mathbf{k}$ and the point B has position vector $5\mathbf{i} + 2\mathbf{j} + 6\mathbf{k}$. The line l passes through the points A and B.

a Find the vector $\overrightarrow{AB}$. **(2 marks)**

b Find a vector equation for the line l. **(2 marks)**

The point C has position vector $4\mathbf{i} + 10\mathbf{j} + 2\mathbf{k}$

The point P lies on l. Given that the vector $\overrightarrow{CP}$ is perpendicular to l,

c find the position vector of the point P. **(6 marks)**

(E/P) **32** With respect to a fixed origin O, the lines l_1 and l_2 are given by the equations

$$l_1 : \mathbf{r} = \begin{pmatrix} 3 \\ -2 \\ 4 \end{pmatrix} + \lambda \begin{pmatrix} 2 \\ 1 \\ -1 \end{pmatrix} \qquad l_2 : \mathbf{r} = \begin{pmatrix} 1 \\ 12 \\ 8 \end{pmatrix} + \mu \begin{pmatrix} 1 \\ -2 \\ -1 \end{pmatrix}$$

where λ and μ are scalar parameters.

a Show that l_1 and l_2 meet and find the position vector of their point of intersection, A.   **(6 marks)**

b Find, to the nearest $0.1°$, the acute angle between l_1 and l_2.   **(3 marks)**

The point B has position vector $\begin{pmatrix} 5 \\ -1 \\ 3 \end{pmatrix}$

c Show that B lies on l_1.   **(1 mark)**

d Find the shortest distance from B to the line l_2, giving your answer to 3 significant figures.   **(4 marks)**

(E/P) **33** Two aeroplanes are modelled as travelling in straight lines. Aeroplane A travels from a point with position vector $\begin{pmatrix} 120 \\ -80 \\ 13 \end{pmatrix}$ km to a point with position vector $\begin{pmatrix} 200 \\ 20 \\ 5 \end{pmatrix}$ km, relative to a fixed origin O. Aeroplane B starts at a point with position vector $\begin{pmatrix} -20 \\ 35 \\ 5 \end{pmatrix}$ km relative to O, and flies in the direction of $\begin{pmatrix} 10 \\ -2 \\ 0.1 \end{pmatrix}$

a Show that the flight paths of the two aeroplanes will intersect, and determine the position vector of the point of intersection.   **(7 marks)**

An air traffic controller states that this means that the planes will collide.

b Explain why this conclusion is not necessarily correct.   **(2 marks)**

Summary of key points

1 If $\overrightarrow{PQ} = \overrightarrow{RS}$ then the line segments PQ and RS are equal in length and are parallel.

2 $\overrightarrow{AB} = -\overrightarrow{BA}$ as the line segment AB is equal in length, parallel and in the opposite direction to BA.

3 **Triangle law for vector addition:** $\overrightarrow{AB} + \overrightarrow{BC} = \overrightarrow{AC}$

   If $\overrightarrow{AB} = \mathbf{a}$, $\overrightarrow{BC} = \mathbf{b}$ and $\overrightarrow{AC} = \mathbf{c}$, then $\mathbf{a} + \mathbf{b} = \mathbf{c}$

4 Subtracting a vector is equivalent to 'adding a negative vector': $\mathbf{a} - \mathbf{b} = \mathbf{a} + (-\mathbf{b})$

5 Adding the vectors $\overrightarrow{PQ}$ and $\overrightarrow{QP}$ gives the zero vector $\mathbf{0}$: $\overrightarrow{PQ} + \overrightarrow{QP} = \mathbf{0}$

6 Any vector parallel to the vector $\mathbf{a}$ may be written as $\lambda\mathbf{a}$, where λ is a non-zero scalar.

7 To multiply a column vector by a scalar, multiply each component by the scalar: $\lambda\begin{pmatrix} p \\ q \end{pmatrix} = \begin{pmatrix} \lambda p \\ \lambda q \end{pmatrix}$

8 To add two column vectors, add the x-components and the y-components $\begin{pmatrix} p \\ q \end{pmatrix} + \begin{pmatrix} r \\ s \end{pmatrix} = \begin{pmatrix} p + r \\ q + s \end{pmatrix}$

9 A unit vector is a vector of length 1. The unit vectors along the x- and y-axes are usually denoted by $\mathbf{i}$ and $\mathbf{j}$ respectively. $\mathbf{i} = \begin{pmatrix} 1 \\ 0 \end{pmatrix}$ $\mathbf{j} = \begin{pmatrix} 0 \\ 1 \end{pmatrix}$

10 For any two-dimensional vector: $\begin{pmatrix} p \\ q \end{pmatrix} = p\mathbf{i} + q\mathbf{j}$

11 For the vector $\mathbf{a} = x\mathbf{i} + y\mathbf{j} = \begin{pmatrix} x \\ y \end{pmatrix}$, the magnitude of the vector is given by: $|\mathbf{a}| = \sqrt{x^2 + y^2}$

12 A unit vector in the direction of $\mathbf{a}$ is $\dfrac{\mathbf{a}}{|\mathbf{a}|}$

13 In general, a point P with coordinates (p, q) has position vector: $\overrightarrow{OP} = p\mathbf{i} + q\mathbf{j} = \begin{pmatrix} p \\ q \end{pmatrix}$

14 $\overrightarrow{AB} = \overrightarrow{OB} - \overrightarrow{OA}$, where $\overrightarrow{OA}$ and $\overrightarrow{OB}$ are the position vectors of A and B respectively.

15 If the point P divides the line segment AB in the ratio $\lambda : \mu$, then

$$\overrightarrow{OP} = \overrightarrow{OA} + \frac{\lambda}{\lambda + \mu} \overrightarrow{AB}$$

$$= \overrightarrow{OA} + \frac{\lambda}{\lambda + \mu} (\overrightarrow{OB} - \overrightarrow{OA})$$

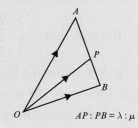

$AP : PB = \lambda : \mu$

16 If $\mathbf{a}$ and $\mathbf{b}$ are two non-parallel vectors and $p\mathbf{a} + q\mathbf{b} = r\mathbf{a} + s\mathbf{b}$ then $p = r$ and $q = s$

17 The distance from the origin to the point (x, y, z) is $\sqrt{x^2 + y^2 + z^2}$

18 The distance between the points (x_1, y_1, z_1) and (x_2, y_2, z_2) is $\sqrt{(x_1 - x_2)^2 + (y_1 - y_2)^2 + (z_1 - z_2)^2}$

19 The unit vectors along the x-, y- and z-axes are denoted by $\mathbf{i}$, $\mathbf{j}$ and $\mathbf{k}$ respectively.

$\mathbf{i} = \begin{pmatrix} 1 \\ 0 \\ 0 \end{pmatrix}$ $\qquad \mathbf{j} = \begin{pmatrix} 0 \\ 1 \\ 0 \end{pmatrix}$ $\qquad \mathbf{k} = \begin{pmatrix} 0 \\ 0 \\ 1 \end{pmatrix}$

Any 3D vector can be written in column form as $p\mathbf{i} + q\mathbf{j} + r\mathbf{k} = \begin{pmatrix} p \\ q \\ r \end{pmatrix}$

20 If the vector $\mathbf{a} = x\mathbf{i} + y\mathbf{j} + z\mathbf{k}$ makes an angle θ_x with the positive x-axis then $\cos\theta_x = \dfrac{x}{|\mathbf{a}|}$ and similarly for the angles θ_y and θ_z.

21 If $\mathbf{a}$, $\mathbf{b}$ and $\mathbf{c}$ are vectors in three dimensions which do not all lie in the same plane then you can compare their coefficients on both sides of an equation.

22 A vector equation of a straight line passing through the point A with position vector **a**, and parallel to the vector **b**, is

$\quad$ **r** = **a** + λ**b** where λ is a scalar parameter.

23 A vector equation of a straight line passing through the points C and D, with position vectors **c** and **d** respectively, is

$\quad$ **r** = **c** + λ(**d** − **c**) where λ is a scalar parameter.

24 The **scalar product** of two vectors **a** and **b** is written as **a.b** (say 'a dot b'), and defined as

$\quad$ **a.b** = |**a**||**b**|$\cos\theta$

where θ is the angle between **a** and **b**.

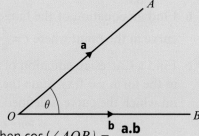

25 If **a** and **b** are the position vectors of the points A and B, then $\cos(\angle AOB) = \dfrac{\textbf{a.b}}{|\textbf{a}||\textbf{b}|}$

26 The non-zero vectors **a** and **b** are perpendicular if and only if **a.b** = 0.

27 If **a** and **b** are parallel, **a.b** = |**a**||**b**| In particular, **a.a** = |**a**|²

28 If **a** = a_1**i** + a_2**j** + a_3**k** and **b** = b_1**i** + b_2**j** + b_3**k**

$$\textbf{a.b} = \begin{pmatrix} a_1 \\ a_2 \\ a_3 \end{pmatrix} . \begin{pmatrix} b_1 \\ b_2 \\ b_3 \end{pmatrix} = a_1 b_1 + a_2 b_2 + a_3 b_3$$

29 The acute angle θ between two intersecting straight lines is given by:

$\quad \cos\theta = \left| \dfrac{\textbf{a.b}}{|\textbf{a}||\textbf{b}|} \right|$ where **a** and **b** are direction vectors of the lines.

30 Two lines are **skew** if they are not parallel and they do not intersect.

Review exercise

2

(E) 1 A curve has parametric equations
$$x = 2\cot t, \ y = 2\sin^2 t, \ 0 < t \leqslant \frac{\pi}{2}$$
a Find $\dfrac{dy}{dx}$ in terms of t. **(3)**

b Find an equation of the tangent to the curve at the point where $t = \dfrac{\pi}{4}$ **(3)**

c Find a Cartesian equation of the curve in the form $y = f(x)$. State the domain on which the curve is defined. **(3)**

← **Pure 4 Section 5.1**

(E/P) 2 The curve C has parametric equations
$$x = \frac{1}{1+t}, \ y = \frac{1}{1-t}, \ -1 < t < 1$$

The line l is a tangent to C at the point where $t = \dfrac{1}{2}$

a Find an equation for the line l. **(5)**

b Show that a Cartesian equation for the curve C is $y = \dfrac{x}{2x-1}$ **(3)**

← **Pure 4 Section 5.1**

(E/P) 3 A curve C is described by the equation
$$3x^2 - 2y^2 + 2x - 3y + 5 = 0$$

Find an equation of the normal to C at the point $(0, 1)$, giving your answer in the form $ax + by + c = 0$, where a, b and c are integers. **(7)**

← **Pure 4 Section 5.2**

(E/P) 4 A set of curves is given by the equation
$$\sin x + \cos y = 0.5$$

a Use implicit differentiation to find an expression for $\dfrac{dy}{dx}$ **(4)**

For $-\pi < x < \pi$ and $-\pi < y < \pi$

b find the coordinates of the points where $\dfrac{dy}{dx} = 0$ **(3)**

← **Pure 4 Section 5.2**

(E/P) 5 The volume of a spherical balloon of radius r cm is V cm³, where $V = \dfrac{4}{3}\pi r^3$

a Find $\dfrac{dV}{dr}$ **(1)**

The volume of the balloon increases with time t seconds according to the formula
$$\frac{dV}{dt} = \frac{1000}{(2t+1)^2}, \ t \geqslant 0$$

b Find an expression in terms of r and t for $\dfrac{dr}{dt}$ **(3)**

← **Pure 4 Section 5.3**

(E) 6 A curve is given as $x^3 + 3x^2y = 4$. The point $P(1,1)$ lies on the curve.

a Find $\dfrac{dy}{dx}$ in terms of x and y.

b Find the tangent to the curve at the point P.

← **Pure 4 Section 6.1**

(E) 7 The curve C is represented by the parametric equations
$$x = \ln(2t-1), \ y = at - 3t^3 \text{ for } t > k$$

a State the minimum value of k, where k is a rational number.

b Given that $t = \dfrac{2}{3}$ when the gradient is 0, determine the value of a.

c Find the equation of the normal to the curve when $t = 1$

← **Pure 4 Section 6.1**

8 A curve is represented by the parametric equations $x = \sec^2 t$, $y = \cot t$ for the interval $0 \leq t \leq \dfrac{\pi}{4}$

a Show that $\dfrac{dx}{dt} = k\sec^2 t \tan t$, where k is to be determined.

b Determine the area under the curve over the given interval. Give your answer in an exact form.

← **Pure 4 Section 6.1**

9 The curve C is represented by the parametric equations $x = \frac{2}{3}t^{\frac{3}{2}}$, $y = 2t^{\frac{5}{2}}$

Given that the area under the curve from $t = a$ to $t = 3$ is 40, determine the exact value of a, where $a > 0$

← **Pure 4 Section 6.1**

(E) **10** The curve shown in the diagram is $y = x\sqrt{1 - x^2}$

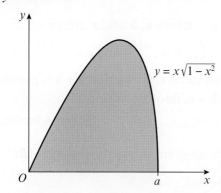

a Write down the value of a. **(1)**

The finite shaded region bounded by the curve and the x-axis is rotated through 2π radians about the x-axis.

b Find the exact volume of the solid generated. **(5)**

← **Pure 4 Section 6.1**

(E) **11** A curve is represented by the parametric equations $x = \tan t$, $y = \cos^2 t$

This curve is then rotated about the x-axis to generate a volume between the values $t = 0$ and $t = \frac{\pi}{4}$

Determine the exact volume generated.

← **Pure 4 Section 6.1**

(E) **12** Using the substitution $u^2 = 2x - 1$ or otherwise, find the exact value of

$$\int_1^5 \frac{3x}{\sqrt{2x - 1}}\,dx \qquad (6)$$

← **Pure 4 Section 6.2**

(E) **13** Use the substitution $u = 1 - x^2$ to find the exact value of

$$\int_0^{\frac{1}{2}} \frac{x^3}{(1 - x^2)^{\frac{1}{2}}}\,dx \qquad (6)$$

← **Pure 4 Section 6.2**

(E/P) **14** $f(x) = (x^2 + 1)\ln x$

Find the exact value of $\int_1^e f(x)\,dx$ **(7)**

← **Pure 4 Section 6.3**

(E/P) **15 a** Express $\dfrac{5x + 3}{(2x - 3)(x + 2)}$ in partial fractions. **(3)**

b Hence find the exact value of

$\int_2^6 \dfrac{5x + 3}{(2x - 3)(x + 2)}\,dx$, giving your answer as a single logarithm. **(4)**

← **Pure 4 Section 6.4**

(E/P) **16** Integrate $\int e^{-x}\cos 2x\,dx$.

← **Pure 4 Section 6.4**

(E/P) **17 a** Express $\dfrac{2x - 1}{(x - 1)(2x - 3)}$ in partial fractions. **(4)**

b Given that $x \geqslant 2$, find the general solution of the differential equation

$$(2x - 3)(x - 1)\frac{dy}{dx} = (2x - 1)y \qquad (4)$$

c Hence find the particular solution of this differential equation that satisfies $y = 10$ at $x = 2$ giving your answer in the form $y = f(x)$ **(2)**

← **Pure 4 Sections 6.4, 6.5**

(E/P) **18** A spherical balloon is being inflated in such a way that the rate of increase of its volume, $V\,\text{cm}^3$, with respect to time t seconds is given by $\dfrac{dV}{dt} = \dfrac{k}{V}$ where k is a positive constant.

Given that the radius of the balloon is r cm, and that $V = \frac{4}{3}\pi r^3$

a prove that r satisfies the differential equation

$$\frac{dr}{dt} = \frac{B}{r^5}$$

where B is a constant. **(4)**

b Find a general solution of the differential equation obtained in part **a**.   **(5)**

← **Pure 4 Sections 5.3, 6.5, 6.6**

(E/P) 19 Liquid is pouring into a container at a constant rate of $20\,\text{cm}^3\,\text{s}^{-1}$ and is leaking out at a rate proportional to the volume of the liquid already in the container.

a Explain why, at time t seconds, the volume, $V\,\text{cm}^3$, of liquid in the container satisfies the differential equation

$$\frac{dV}{dt} = 20 - kV$$

where k is a positive constant.   **(2)**

The container is initially empty.

b By solving the differential equation, show that

$$V = A + Be^{-kt}$$

giving the values of A and B in terms of k.   **(5)**

Given also that $\dfrac{dV}{dt} = 10$ when $t = 5$

c find the volume of liquid in the container at $10\,\text{s}$ after the start.   **(3)**

← **Pure 4 Sections 6.5, 6.6**

(E/P) 20 The rate of decrease of the concentration of some medicine in the blood stream is proportional to the concentration C of the medicine which is present at that time. The time t is measured in hours from the administration of the medicine (i.e. when it was given) and C is measured in micrograms per litre.

a Show that this process is described by the differential equation $\dfrac{dC}{dt} = -kC$ explaining why k is a positive constant.   **(2)**

b Find the general solution of the differential equation, in the form $C = f(t)$ **(4)**

After 4 hours, the concentration of the medicine in the bloodstream is reduced to 10% of its starting value C_0

c Find the exact value of k.   **(3)**

← **Pure 4 Sections 6.5, 6.6**

(E) 21 The vector $9\mathbf{i} + q\mathbf{j}$ is parallel to the vector $2\mathbf{i} - \mathbf{j}$. Find the value of the constant q.   **(2)**

← **Pure 4 Sections 7.2**

(E/P) 22 Given that $|5\mathbf{i} - k\mathbf{j}| = |2k\mathbf{i} + 2\mathbf{j}|$, find the exact value of the positive constant k.   **(4)**

← **Pure 4 Sections 7.3**

(E/P) 23 The vectors **a**, **b** and **c** are given as
$\mathbf{a} = \begin{pmatrix} 8 \\ 23 \end{pmatrix}$, $\mathbf{b} = \begin{pmatrix} -15 \\ x \end{pmatrix}$ and $\mathbf{c} = \begin{pmatrix} -13 \\ 2 \end{pmatrix}$, where x is an integer. Given that $\mathbf{a} + \mathbf{b}$ is parallel to $\mathbf{b} - \mathbf{c}$, find the value of x.   **(4)**

← **Pure 4 Sections 7.2**

(E/P) 24 The diagram shows the triangle ABC.

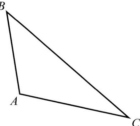

Given that $\overrightarrow{AB} = -\mathbf{i} + 6\mathbf{j} + 4\mathbf{k}$ and $\overrightarrow{AC} = 5\mathbf{i} - 2\mathbf{j} - 3\mathbf{k}$ find the size of $\angle BAC$ to one decimal place.   **(5)**

← **Pure 4 Sections 7.4**

(E) 25 P is the point $(-6, 3, 2)$ and Q is the point $(4, -2, 0)$. Find

a the vector $\overrightarrow{PQ}$ **(1)**

b the unit vector in the direction of $\overrightarrow{PQ}$ **(2)**

c the angle $\overrightarrow{PQ}$ makes with the positive z-axis.   **(2)**

The vector $\overrightarrow{AB} = 30\mathbf{i} - 15\mathbf{j} + 6\mathbf{k}$

d Explain, with a reason, whether the vectors $\overrightarrow{AB}$ and $\overrightarrow{PQ}$ are parallel. **(2)**

← **Pure 4 Sections 7.4**

(E/P) **26** The vertices of triangle MNP have coordinates $M(-2, 0, 5)$, $N(8, -5, 1)$ and $P(k, -2, -6)$. Given that triangle MNP is isosceles and k is a positive integer, find the value of k. **(4)**

← **Pure 4 Sections 7.6**

(E/P) **27** Given that
$$-6\mathbf{i} + 40\mathbf{j} + 16\mathbf{k} = 3p\mathbf{i} + (8 + qr)\mathbf{j} + 2pr\mathbf{k}$$
find the values of p, q and r. **(3)**

← **Pure 4 Sections 7.6**

(E) **28** The line l has equation
$$\mathbf{r} = \begin{pmatrix} 2 \\ -1 \\ 3 \end{pmatrix} + \lambda \begin{pmatrix} -1 \\ -2 \\ 3 \end{pmatrix}$$

A and B are the points on l with $\lambda = 4$ and $\lambda = -1$ respectively.

Find the distance AB. **(4)**

← **Pure 4 Sections 7.9**

(E/P) **29** The points $P(1, -1, 3)$, $Q(a, 3, 8)$ and $R(5, 7, b)$, where a and b are constants, are collinear. Find the values of a and b and the vector equation of the line through the three points. **(5)**

← **Pure 4 Sections 7.9**

(E) **30** The line l_1 has vector equation
$$\mathbf{r} = 11\mathbf{i} + 5\mathbf{j} + 6\mathbf{k} + \lambda(4\mathbf{i} + 2\mathbf{j} + 4\mathbf{k})$$
and the line l_2 has vector equation
$$\mathbf{r} = 24\mathbf{i} + 4\mathbf{j} + 13\mathbf{k} + \mu(7\mathbf{i} + \mathbf{j} + 5\mathbf{k})$$
where λ and μ are parameters.

a Show that the lines l_1 and l_2 intersect. **(3)**

b Find the coordinates of their point of intersection. **(2)**

Given that θ is the acute angle between l_1 and l_2,

c find the value of $\cos\theta$. Give your answer in the form $k\sqrt{3}$, where k is a simplified fraction. **(3)**

← **Pure 4 Sections 7.10, 7.11**

(E/P) **31** The line l_1 has vector equation
$$\mathbf{r} = 8\mathbf{i} + 12\mathbf{j} + 14\mathbf{k} + \lambda(\mathbf{i} + \mathbf{j} - \mathbf{k})$$
The points A, with coordinates $(4, 8, a)$ and B, with coordinates $(b, 13, 13)$, lie on this line.

a Find the values of a and b. **(2)**

Given that the point O is the origin, and that the point P lies on l_1 such that OP is perpendicular to l_1,

b find the coordinates of P. **(3)**

c Hence find the distance OP, giving your answer as a simplified surd. **(3)**

← **Pure 4 Sections 7.10, 7.11**

(E/P) **32** Two fish are modelled as travelling in straight lines. A shark, swims from a point with position vector $\begin{pmatrix} 2 \\ 3 \\ -1 \end{pmatrix}$ to a point with position vector $\begin{pmatrix} -2 \\ 11 \\ 11 \end{pmatrix}$ both relative to a fixed origin O and with units given in metres.

A flounder, starts at a point with position vector $\begin{pmatrix} 2 \\ 0 \\ 1 \end{pmatrix}$ relative to O, and travels in the direction of $\begin{pmatrix} -2 \\ -1 \\ 3 \end{pmatrix}$.

a Show that, no matter how fast either fish swims, the shark will never catch the flounder. **(7)**

b Give one criticism of this model. **(1)**

← **Pure 4 Sections 7.10**

Challenge

1 A curve C is described as $x^3 - xy^2 = y + 5$.

a Find an expression for $\dfrac{dy}{dx}$ in terms of x and y.

b Find an expression for $\dfrac{d^2y}{dx^2}$ in terms of x, y and $\dfrac{dy}{dx}$

c Hence find the values of $\dfrac{dy}{dx}$ and $\dfrac{d^2y}{dx^2}$ at the point , giving your answers as fractions.

2 By considering a suitable substitution, determine the value of $\displaystyle\int_{e^2}^{e^3} \dfrac{1}{x(\ln x)}\,dx$

Give your answer in an exact form.

3 The points A, B and C have coordinates $(-2, -3, 0)$, $(-1, -1, 3)$ and $(1, 1, 1)$ respectively. Find the centre and radius of the circle that passes through all three points.

Hint For question **3**, consider angle ABC.

Exam practice

Further Mathematics International Advanced Level
Pure Mathematics 4

Time: 1 hour 30 minutes
You must have: Mathematical Formulae and Statistical Tables, Calculator

1   Prove by contradiction that if $n^2 + 1$ is even then n must be odd.   **(4)**

2   A curve is given as $x^2 + 4xy + y^2 + 1 = 0$

   a Find $\dfrac{dy}{dx}$ in terms of x and y.   **(5)**

   b Determine the coordinates where the curve is parallel to the x-axis.   **(5)**

3   A curve C is given as $y = xe^{2x}$

   a Find the exact coordinates of the turning point.   **(4)**

   b Find the volume generated when the curve is rotated about the x-axis
     through 2π radians, between the values $x = 1$ and $x = 2$ **(7)**

4   A curve has parametric equations

$$x = \cos t \qquad y = \cos 2t \qquad 0 \leq t \leq 2\pi$$

   a Show that $y = ax^2 + b$, where a, b are constants to be determined.   **(3)**

   b State the domain and range of the function.   **(2)**

   c Find $\dfrac{dy}{dx}$ in terms of t, giving your answer in a simplified form.   **(3)**

   d Determine the equation of the tangent at the point on the curve where $t = 0$ **(3)**

5   **a** Use the binomial series to find the expansion of

$$\frac{1}{(1 - 3x)^2} \qquad |x| < \frac{1}{3}$$

     in ascending powers of x, up to and including the term in x^3 **(6)**

   b Hence, find an approximation for $\dfrac{1}{0.97^2}$, giving your answer to 6 decimal places.   **(3)**

6 The height of water, H, in a storage tank is modelled by the differential equation

$$\frac{\mathrm{d}H}{\mathrm{d}t} = -20(H - 5)$$

where t represents the time in hours.

a Given that $H = 40$ when $t = 0$, find $H = \mathrm{f}(t)$ **(8)**

b By finding $\dfrac{\mathrm{d}H}{\mathrm{d}t}$ in terms of t, determine the value of $\dfrac{\mathrm{d}H}{\mathrm{d}t}$ when $t = 0.5$ **(3)**

c State the eventual height of the water in the tank. **(1)**

7 You are given two points $A(2, 1, 3)$ and $B(5, -2, 1)$

a Find the vector equation of the line l that passes through A and B. **(3)**

b Show that the point $(-4, 7, 7)$ is on the line l. **(2)**

c Find the angle between I and the x-axis. **(3)**

d The point C lies on the line l such that $AC = 3AB$

Determine the possible coordinates of the point C. **(4)**

8 The curve C has the set of parametric equations

$$x = t^2 \qquad y = 1 - t$$

The curve is rotated about the x-axis by 2π radians from $t = 0$ to t $= 1$

Find the exact volume generated. **(6)**

TOTAL FOR PAPER: 75 MARKS

GLOSSARY

accurate a value that is precise or correct to a large number of decimal places

acute angle an angle that is less than 90°

algebraic a mathematical **expression** consisting of numbers, operations (add, subtract, etc.) and letters representing unknown values

anchored fixed in position

anticlockwise the opposite **direction** in which the hands of a clock move around

approximation a number that is not **exact**

ascending increasing

axis (plural axes) either of the two lines by which the positions of points are measured in a graph

base area the cross-sectional area of a three-dimensional shape. For example, a pyramid has a square *base area*

bearing the angle measured in a **clockwise** direction from the north **direction**

binomial an **algebraic expression** of the sum or difference of two **terms**
for example; $(a + b)^n$ is the general form of a binomial expression

bisect to cut in two **exact** halves

boundary condition a **condition** that is required to be satisfied at all, or part of, the boundary of a **region** in which a set of **differential equations** is to be solved

bounded enclosed

Cartesian a unique point in two or three-dimensional space specified by numerical **coordinates**

circle the set of all points in a **plane** that are the same distance from a given point; the centre

clockwise the same **direction** in which the hands of a clock move around

coefficient a numerical or constant quantity placed before and multiplying the **variable** in an **algebraic expression**

collinear points lying on the same line

column vector a **vector** of the form $\begin{pmatrix} 3 \\ 4 \end{pmatrix}$ or $\begin{pmatrix} 1 \\ -2 \\ 3 \end{pmatrix}$

common factor a quantity that will divide without remainder into two or more other quantities

concave an outline or surface that curves inwards like the interior of a **circle** or **sphere**

condition a rule that must be met

conical a cone shaped object

constant a **term** that does not include a **variable**. In the **expression** $3x^3 - 5x + 4$, the constant term is 4

contradiction a statement that disagrees with another statement

converge to approach a limiting value as, either the value of the **variable increases**, or in the case of a **series**, the number of **terms** increases
for example, $\frac{1}{x}$ *converges* to 0 as x approaches ∞

convert change into something else

convex an outline or surface like the exterior of a circle or sphere

coordinates a set of values that show an **exact** position. In a two-dimensional grid, the first number represents a point on the x-**axis** and the second number represents a point on the y-axis

coplanar points lying in the same **plane**

corresponding an equivalent; connected with what you have just mentioned

cube a three-dimensional object that has all edges of the same length

cubic an **algebraic expression** in which the highest **power** is 3

cuboid a six-sided solid whose sides are all rectangles, placed at right angles. Can be, but not necessarily, a **cube**

curve a function that is not linear in shape. For example, $f(x) = \frac{1}{x}$

cylindrical a cylinder-shaped object

decrease to go down in value

deduce to reach a logical conclusion. If $x - 5 = 2$, we can *deduce* that $x = 7$

definite integral an **integral** that has a numerical solution. For example, $\int_1^2 x \, dx$ is a *definite integral*, whereas $\int x \, dx$ is an *indefinite integral*

degree (of a polynomial) the highest **power** in an **expression determines** the degree of the expression. For example, $7x^3 + 5x^2$ is a **polynomial** of *degree* 3 since this is the highest power

degrees an angular measurement where 360° describes a full circle

denominator the lower part of a **fraction**. b in $\frac{a}{b}$

derivative a way to represent the **rate of change** of a mathematical function

determine to find out. By analysing the **equation** of a **curve**, we can *determine* whether or not it is a circle

diagonal a line that joins two non-adjacent vertices in a polygon.

differential equation an **equation** that contains a **derivative**. For example, $\frac{dy}{dx} = x + y$ is a first order *differential equation* since it contains a 1st derivative **term**

differentiate to **determine** the **derivative** of a function. To find its **rate of change** with respect to another **variable**

differentiation the instantaneous **rate of change** of a function with respect to one of its variables

dimensions the size of something, usually given as its length, width and height

direction a course along which something moves. For example, north, east, etc. are **directions**. **Vectors** have direction

displacement change of position

distinct different or unique

domain the set of values that a function exists for. For example, $f(x) = \sqrt{x}$ only exists for $x \geq 0$

eliminate to remove a **variable** from a set of **equations**

endpoints a point at the start or end of an interval, a line **segment**, or part of a **curve**

equate put equal to each other, usually to **determine** a relationship or solution. For example, equating x^2 and x yields solutions of 0 and 1

equation a statement that values of two mathematical **expressions** are equal

error a measure of the difference between an **exact** value and an **estimated** value

estimate roughly calculate a value

evaluate calculate the value of. If $x = 2$, we can evaluate $4x$ to 8

even number an **integer** that is divisible by 2 without leaving a remainder

exact not approximated in any way, precise

expand to multiply out **terms** in brackets

expansion a mathematical expression written in an extended form
for example, the *expansion* of $(a + b)^2$ is $a^2 + 2ab + b^2$

exponential something is said to **increase** or **decrease** *exponentially* if its **rate of change** is **expressed** using exponents. Functions that are *exponential* are of the form $y = a \times b^x$, where a, b are **constants**, and b is the base

expression any group of **terms** that represents something mathematical
$x + 2$, $3x^4$ and $xy + y^3$ are expressions

factor a number that divides into another number exactly

factorise to rewrite an **expression** using brackets
ee *factorise* $x^2 + 3x$ to get $x(x + 3)$

finite a value that is limited; not **infinite**

formula (plural formulae) a mathematical **expression** or **equation** with a particular meaning

fraction a mathematical **expression** representing the division of one whole number by another

general solution a solution that contains unknown **constants**. For example, from **integration** $\int 2x \, dx$ produces the *general solution* $x^2 + c$, where the constant of integration is unknown

generate produce, create

geometry a part of mathematics relating to points, lines, areas and solids

gradient slope

hemisphere half a **sphere**

hexagon a six-sided two-dimensional shape

identical exactly the same as something else

identity an equality between **expressions** that is true for all values of the **variables** in those expressions

implicit a function that is not expressed directly in **terms** of independent **variables**

improper fraction a **fraction** that is represented in the form $\frac{5}{2}$ where the **numerator** is greater than the **denominator**. Also known as **top-heavy fractions**

increase to go up in value

index (plural indices) the **power** to which a quantity is raised, shown as a small number (superscript)

infinite a value that is unlimited, not **finite**

inflection point a point on a **curve** at which a change in the **direction** of curvature occurs

insignificant too small or unimportant to be worth consideration

integer a whole number (1,2, etc.)

integral a function of which a given function is the **derivative**

integrand a function that is to be integrated

integration one of the two operations of calculus, the other being **differentiation**, its **inverse**

intersection the point at which two or more **curves** cross (intersect)

inverse operations that reverse each other. For example, the *inverse* of $f(x) = x^3$ is $f^{-1}(x) = x^{\frac{1}{3}}$

inversely proportional when the **product** of two **variables** equals a **constant**, these variables are said to be *inversely proportional* to each other. For example, $y = \dfrac{k}{x}$ shows that y is *inversely proportional* to x

irrational number a real number that cannot be expressed as a fraction or an **integer**

isosceles a **triangle** with two equal sides, hence two equal angles

limit a point or value that a sequence, or function, or sum of a **series** will approach, until it is as close to the point or value as desired

linear functions that have **variables** with **power** 1. For example, $y = 4x - 3$

logarithm the **power** to which the base number must be raised in order to get a particular number. For example, $\log_2 32 = 5 \Rightarrow 2^5 = 32$

magnitude a numerical quantity or value, taken as **positive**. For example, the *magnitude* of -3 is 3

midpoint a point which lies directly between two other points

model an attempt to describe, explore and solve real-life systems and problems using mathematical concepts and language

modulus the magnitude or absolute value of an expression or constant.

multiple a number that can be divided into another number without a remainder. For example, 8 is a *multiple* of 32

natural logarithm a **logarithm** that has base e. For example, $\log_e 5$, which is usually written as ln 5

negation a **contradiction** of something

negative a value that is less than zero

non-parallel two lines or edges that are not the same distance from each other at all points

non-zero a value that is not equal to zero

normal a line that is **perpendicular** (at right angles) to another line. For example, the *normal* to a **curve** is perpendicular to the tangent to the curve

numerator the upper part of a fraction. a in $\dfrac{a}{b}$

obtain get, or acquire. For example, putting $x = 2$ into $y = 4x$ allows you to *obtain* the value for y

odd an **integer** that is not a **multiple** of 2

origin the point where the x-**axis** and y-axis intersect on a flat coordinate **plane**

parallel two lines side-by-side, the same distance apart at every point

parallelepiped a solid body that has a **parallelogram** for each face

parallelogram a four-sided shape that has two pairs of **parallel** opposite sides

parametric equations a set of **equations** that express a set of quantities as functions of a number of independent **variables**, known as "parameters". For example, $x = t^2$, $y = 2t$, $t \in \mathbb{R}$ are a set of *parametric equations*

particular solution a solution to a **differential equation** that contains no unknown **constants** of **integration**

percentage a part of a whole expressed in hundredths

perpendicular means at right angles. A line meeting another line at 90°

plane a flat surface on which a straight line joining any two points on it would lie

plot to make a diagram, chart or curve, for example, using certain values

polynomial an **expression** of two or more **algebraic terms** with **positive** whole number indices. For example, $6x^4 - 7x^2$ is a *polynomial*

position vector a **vector** that starts from the **origin**.

positive a value that is greater than zero

power if a number is **increased** to the *power* of three, then it is multiplied by itself three times

For example, x to the *power* of 2 is $x \times x$, written as x^2

prime number a number that can only be divided by itself and 1

product the *product* of two numbers is the result of multiplying them together. For example, $3 \times 4 = 12$ so 12 is the *product* of 3 and 4

proportional a pair of **variables** that have a **constant** ratio to each other. For example, $y = kx$ shows that $\dfrac{y}{x}$ is always the same amount

quadratic an **expression** of the form $ax^2 + bx + c$ where $a \neq 0$

quadrilateral a four-sided shape

quotient a result obtained by dividing one quantity by another

radian an angle subtended by a circular arc as the length of the arc divided by the **radius** of the arc. One radian is approximately 57.30°

radius the line **segment** joining the centre to the circumference of a circle

rate of change the change in one **variable** with respect to the change in another variable

rational any real number, including **integers**, that can be written as a fraction, even if it produces a repeating decimal

rearrange to put **terms** in a different order

region an area of a graph enclosed by **curves** or lines

repeated factor an **expression**, or value, that contains the same **factor** more than once. For example, in the expression $(x + 2)(x - 1)^2$, $x - 1$ is a *repeated factor*

resultant a **vector** quantity which is equivalent to the combined effect of two or more other vectors acting at the same point

revolution one *revolution* is a complete turn through 360°

right-angle an angle between two lines where the angle is 90°. This is also known as **perpendicular**

root a solution of a quadratic equation

rotate turn about a specified point

scalar a real number that is not a **vector**

scalene a **triangle** that has no equal sides or equal angles

segment a smaller part of a larger object. For example, a line segment is a line of **finite** length between two points

series the sum of **terms** in a sequence

significant figures digits counted from the first (**non-zero**) digit on the left

simplify to replace an **expression** with a simpler, usually shorter, one

sketch a drawing that explains something without necessarily being **accurate**

skew lines two lines that are not **parallel** and do not intersect

sphere a solid three-dimensional shape which has every point on its outer surface the same distance from the centre

square (power) to raise a number or **variable** to the **power** of 2. For example, x squared is written as x^2

square root (of) a number which, when multiplied by itself, equals x. Can be written as $\sqrt{x}$ or $x^{\frac{1}{2}}$

stationary point A point on a **curve** where the **gradient** is zero

strip a thin rectangular area, used to approximate the area of a **region**, such as in **integration**

substitution the replacing of a **variable** in an **expression** with a value or another representation

sufficiently small a value so small that it can be ignored.. For example, 0.001^4 could usually be considered small enough to ignore

surds numbers that are written in the form $\sqrt{a}$ but cannot be resolved into a **rational** number. For example, $\sqrt{2}$

surface area the total area of the surface of a three-dimensional object

tangent a line that touches a **curve** at a point. The **gradient** of the tangent and the curve are equal at that point

term a separate part of a mathematical **expression**. For example, the expression $2x^4 - 5$ has two terms

top heavy a fraction that has a greater numerator than denominator. For example, $\frac{100}{52}$ and $\frac{x^2}{2x - 3}$ are both top heavy fractions

trapezium a four-sided shape that has one set of opposite sides **parallel** to each other.

triangle a three-sided shape

trigonometric function a function that relates **triangles** and their sides and angles. For example, $\sin x$, $\cos x$ are both trigonometric functions

trisect to cut into three equal parts

unit vector a **vector** that has **magnitude** 1

variable a quantity that is not **constant**

vector a value that has both *magnitude* and **direction**

velocity measured as the **displacement** covered divided by the time taken

vertex (plural vertices) a point where two lines, or edges, meet. Is also used to define the turning point of a **quadratic** function

ANSWERS

CHAPTER 1

Prior knowledge check

1 a $(x-1)(x-5)$ **b** $(x+4)(x-4)$ **c** $(3x-5)(3x+5)$
2 a even **b** either **c** either **d** odd

Exercise 1A

1 B At least one multiple of three is odd.
2 a At least one rich person is not happy.
  b There is at least one prime number between 10 million and 11 million.
  c If p and q are prime numbers there exists a number of the form $(pq + 1)$ that is not prime.
  d There is a number of the form $2^n - 1$ that is neither a prime nor a multiple of 3.
  e None of the above statements are true.
3 a There exists a number n such that n^2 is odd but n is even.
  b n is even so write $n = 2k$
    $n^2 = (2k)^2 = 4k^2 = 2(2k^2) \Rightarrow n^2$ is even.
    This contradicts the assumption that n^2 is odd.
    Therefore if n^2 is odd then n must be odd.
4 a Assumption: there is a greatest even integer $2n$.
    $2(n + 1)$ is also an integer and $2(n + 1) > 2n$
    $2n + 2 = $ even + even = even
    So there exists an even integer greater that $2n$.
    This contradicts the assumption.
    Therefore there is no greatest even integer.
  b Assumption: there exists a number n such that n^3 is even but n is odd.
    n is odd so write $n = 2k + 1$
    $n^3 = (2k + 1)^3 = 8k^3 + 12k^2 + 6k + 1$
    $= 2(4k^3 + 6k^2 + 3k) + 1 \Rightarrow n^3$ is odd.
    This contradicts the assumption that n^3 is even.
    Therefore, if n^3 is even then n must be even.
  c Assumption: if pq is even then neither p nor q is even.
    p is odd, $p = 2k + 1$
    q is odd, $q = 2m + 1$
    $pq = (2k + 1)(2m + 1) = 2km + 2k + 2m + 1$
    $= 2(km + k + m) + 1 \Rightarrow pq$ is odd.
    This contradicts the assumption that pq is even.
    Therefore, if pq is even then at least one of p and q is even.
  d Assumption: if $p + q$ is odd than neither p nor q is odd
    p is even, $p = 2k$
    q is even, $q = 2m$
    $p + q = 2k + 2m = 2(k + m) \Rightarrow$ so $p + q$ is even
    This contradicts the assumption that $p + q$ is odd.
    Therefore, if $p + q$ is odd that at least one of p and q is odd.
5 a Assumption: if ab is an irrational number then neither a nor b is irrational.
    a is rational, $a = \dfrac{c}{d}$ where c and d are integers.
    b is rational, $b = \dfrac{e}{f}$ where e and f are integers.
    $ab = \dfrac{ce}{df}$, ce is an integer, df is an integer.
    Therefore ab is a rational number.

This contradicts assumption that ab is irrational.
Therefore if ab is an irrational number that at least one of a and b is an irrational number.
  b Assumption: if $a + b$ is an irrational number then neither a nor b is irrational.
    a is rational, $a = \dfrac{c}{d}$ where c and d are integers.
    b is rational, $b = \dfrac{e}{f}$ where e and f are integers.
    $a + b = \dfrac{cf + de}{df}$
    cf, de and df are integers.
    So $a + b$ is rational. This contradicts the assumption that $a + b$ is irrational.
    Therefore if $a + b$ is irrational then at least one of a and b is irrational.
  c Many possible answers e.g. $a = 2 - \sqrt{2}$, $b = \sqrt{2}$.
6 Assumption: there exists integers a and b such that $21a + 14b = 1$.
  Since 21 and 14 are multiples of 7, divide both sides by 7.
  So now $3a + 2b = \dfrac{1}{7}$
  $3a$ is also an integer. $2b$ is also an integer.
  The sum of two integers will always be an integer, so $3a + 2b = $ 'an integer'.
  This contradicts the statement that $3a + 2b = \dfrac{1}{7}$
  Therefore there exists no integers a and b for which $21a + 14b = 1$
7 a Assumption: There exists a number n such that n^2 is a multiple of 3, but n is not a multiple of 3.
    We know that all multiples of 3 can be written in the form $n = 3k$, therefore $3k + 1$ and $3k + 2$ are not multiples of 3.
    Let $n = 3k + 1$
    $n^2 = (3k + 1)^2 = 9k^2 + 6k + 1 = 3(3k^2 + 2k) + 1$
    In this case n^2 is *not* a multiple of 3.
    Let $m = 3k + 2$
    $m^2 = (3k + 2)^2 = 9k^2 + 12k + 4 = 3(3k^2 + 4k + 1) + 1$
    In this case n^2 is also not a multiple of 3.
    This contradicts the assumption that n^2 is a multiple of 3.
    Therefore if n^2 is a multiple of 3, n is a multiple of 3.
  b Assumption: $\sqrt{3}$ is a rational number.
    Then $\sqrt{3} = \dfrac{a}{b}$ for some integers a and b.
    Further assume that this fraction is in its simplest terms: there are no common factors between a and b.
    So $3 = \dfrac{a^2}{b^2}$ or $a^2 = 3b^2$
    Therefore a^2 must be a multiple of 3.
    We know from part **a** that this means a must also be a multiple of 3.
    Write $a = 3c$, which means $a^2 = (3c)^2 = 9c^2$
    Now $9c^2 = 3b^2$, or $3c^2 = b^2$
    Therefore b^2 must be a multiple of 3, which implies b is also a multiple of 3.
    If a and b are both multiples of 3, this contradicts the statement that there are no common factors between a and b.
    Therefore, $\sqrt{3}$ is an irrational number.

8 Assumption: there is an integer solution to the equation $x^2 - y^2 = 2$
Remember that $x^2 - y^2 = (x - y)(x + y) = 2$
To make a product of 2 using integers, the possible pairs are: $(2, 1), (1, 2), (-2, -1)$ and $(-1, -2)$.
Consider each possibility in turn.
$x - y = 2$ and $x + y = 1 \Rightarrow x = \frac{3}{2}, y = -\frac{1}{2}$
$x - y = 1$ and $x + y = 2 \Rightarrow x = \frac{3}{2}, y = \frac{1}{2}$
$x - y = -2$ and $x + y = -1 \Rightarrow x = -\frac{3}{2}, y = \frac{1}{2}$
$x - y = -1$ and $x + y = -2 \Rightarrow x = -\frac{3}{2}, y = -\frac{1}{2}$
This contradicts the statement that there is an integer solution to the equation $x^2 - y^2 = 2$
Therefore the original statement must be true: There are no integer solutions to the equation $x^2 - y^2 = 2$.

9 Assumption: $\sqrt[3]{2}$ is rational and can be written in the form $\sqrt[3]{2} = \frac{a}{b}$ and there are no common factors between a and b.
$2 = \frac{a^3}{b^3}$ or $a^3 = 2b^3$
This means that a^3 is even, so a must also be even.
If a is even, $a = 2n$.
So $a^3 = 2b^3$ becomes $(2n)^3 = 2b^3$ which means $8n^3 = 2b^3$ or $4n^3 = b^3$ or $2(2n^3) = b^3$.
This means that b^3 must be even, so b is also even.
If a and b are both even, they will have a common factor of 2.
This contradicts the statement that a and b have common factors.
We can conclude the original statement is true: $\sqrt[3]{2}$ is an irrational number.

10 a m could be non-positive, e.g. if $n = \frac{1}{2}$

b Assumption: There is a least positive rational number, n.
$n = \frac{a}{b}$ where a and b are integers.
Let $m = \frac{a}{2b}$ Since a and b are integer, m is rational and $m < n$.
This contradicts the statement that n is the least positive rational number.
Therefore, there is no least positive rational number.

Chapter review 1

1 a No more than one of the above statements is true
b People in hot countries are happy
c Three quarters of the people that entered the competition did not win a prize
2 Assumption: If ab is rational then either a or b is irrational.
Let $a = \frac{c}{d}$ where c and d are integers.
Let $b = \sqrt{k}$, where k is not a square number.
Then $ab = \frac{c}{d}\sqrt{k}$ which is clearly irrational.
This contradicts the original assumption.
Therefore the original statement must be true: If ab is rational then no single number a or b can be irrational.
3 B No multiples of five are even.

4 Assumption: if $a - 2b$ is irrational then neither a nor b is irrational.
Let $a = \frac{c}{d}$ where c and d are integers.
Let $b = \frac{e}{f}$ where e and f are integers.
Then $a - 2b = \frac{c}{d} - \frac{2e}{f} = \frac{cf - 2de}{df}$ which is clearly rational.
This contradicts the original assumption that $a - 2b$ is irrational.
Therefore the original statement must be true: If $a - 2b$ is irrational then at least one of a and b is irrational.

5 Assumption: There exists integers x and y such that $3x + 18y = 1$
Since 3 is a common factor, $x + 6y = \frac{1}{3}$
Then x and $6y$ are both integers.
The sum $x + 6y$ must also be an integer.
This contradicts the statement that $x + 6y = \frac{1}{3}$
Therefore there exists no integer for which $3x + 18y = 1$

6 Assumption: If n^4 is odd then n can be even.
Let $n = 2k$, where k is an integer.
Then $n^4 = (2k)^4 = 16k^4$ which must be even.
This contradicts the assumption.
Therefore this contradicts the original statement: If n^4 is odd then n must be odd.

CHAPTER 2

Prior knowledge check

1 a $(x + 3)(x + 2)$
b $(x - 7)(x + 2)$
2 a $\frac{x - 3}{x + 6}$
b $-\frac{x + 5}{x + 3}$
3 a $(x - 1)(x^2 - 2x + 3)$
b $(x + 1)(x - 2)^2$

Exercise 2A

1 a $\frac{4}{x + 3} + \frac{2}{x - 2}$ **b** $\frac{3}{x + 1} - \frac{1}{x + 4}$
c $\frac{3}{2x} - \frac{5}{x - 4}$ **d** $\frac{4}{2x + 1} - \frac{1}{x - 3}$
e $\frac{2}{x + 3} + \frac{4}{x - 3}$ **f** $-\frac{2}{x + 1} - \frac{1}{x - 4}$
g $\frac{2}{x} - \frac{3}{x + 4}$ **h** $\frac{3}{x + 5} - \frac{1}{x - 3}$
2 $A = \frac{1}{2}, B = -\frac{3}{2}$
3 $A = 24, B = -2$
4 $A = 1, B = -2, C = 3$
5 $D = -1, E = 2, F = -5$
6 a $\frac{1}{x + 1} - \frac{2}{x - 2} + \frac{3}{x + 5}$
b $-\frac{1}{x} + \frac{2}{2x + 1} - \frac{5}{3x - 2}$
c $\frac{3}{x + 1} - \frac{2}{x + 2} - \frac{6}{x - 5}$

7 a $\dfrac{3}{x} - \dfrac{2}{x+1} + \dfrac{5}{x-1}$

  b $\dfrac{-1}{5x+4} + \dfrac{2}{2x-1}$

Challenge

$\dfrac{6}{x-2} + \dfrac{1}{x+1} - \dfrac{2}{x-3}$

Exercise 2B

1 $A = 0, B = 1, C = 3$ **2** $D = 3, E = -2, F = -4$

3 $P = -2, Q = 4, R = 2$ **4** $C = 3, D = 1, E = 2$

5 $A = 2, B = -4$ **6** $A = 2, B = 4, C = 11$

7 $A = 4, B = 1$ and $C = 12.$

8 a $\dfrac{4}{x+5} - \dfrac{19}{(x+5)^2}$ **b** $\dfrac{2}{x} - \dfrac{1}{2x-1} + \dfrac{6}{(2x-1)^2}$

Exercise 2C

1 a $1 - \dfrac{1}{x+2}$ **b** $-1 + \dfrac{5}{x+1}$

  c $x + 1 + \dfrac{2}{x-1}$ **d** $2x - 6 + \dfrac{18}{x+3}$

2 $1 + \dfrac{4}{5(x-2)} - \dfrac{9}{5(x+3)}$

3 $A = 2, B = 3, C = 1$

4 a $x - 3 + \dfrac{8}{x+3}$ **b** $2 - \dfrac{2}{3x} - \dfrac{16}{3(x+3)}$

  c $-\dfrac{3}{2} + \dfrac{5}{6(2x-1)} + \dfrac{1}{3(x+1)}$

5 $x - 1 + \dfrac{8}{3(x+2)} + \dfrac{1}{3(x-1)}$

6 a $x - 2 + \dfrac{1}{2x} + \dfrac{7}{2(x+2)}$ **b** $x + \dfrac{3}{3(x+2)} + \dfrac{3}{2(x-2)}$

7 $1 - \dfrac{2}{x+1} + \dfrac{1}{(x+1)^2}$

8 a $1 + \dfrac{4}{x-2} + \dfrac{5}{(x+2)^2}$ **b** $2x - 8 \dfrac{24}{x+2} - \dfrac{17}{(x-2)^2}$

Challenge

$3 + \dfrac{9}{x-1} + \dfrac{9}{(x-1)^2} + \dfrac{3}{(x-1)^3}$

Chapter review 2

1 $A = -\dfrac{4}{5}, B = \dfrac{4}{5}$

2 a $\dfrac{8}{x+3} + \dfrac{5}{x+1}$ **b** $\dfrac{1}{3(x-1)} - \dfrac{3}{2(x+3)}$

3 $A = \dfrac{1}{5}, B = \dfrac{2}{21}, C = -\dfrac{5}{28}$

4 a $\dfrac{2}{x} - \dfrac{3}{x+1} + \dfrac{4}{x-1}$ **b** $\dfrac{2}{x-2} + \dfrac{1}{x-2} + \dfrac{3}{x-3}$

  c $\dfrac{1}{2(x+1)} - \dfrac{4}{x+2} + \dfrac{9}{2(x+3)}$

5 $\dfrac{4}{3(x+2)} + \dfrac{5}{21(x-4)} - \dfrac{11}{7(x+3)}$

6 a $-\dfrac{1}{x-1} + \dfrac{1}{3(x+1)} + \dfrac{2}{3(x-2)}$

  b $\dfrac{1}{6x} + \dfrac{3}{2(x+2)} - \dfrac{8}{3(x+3)}$

7 a $-\dfrac{1}{16(x+1)} + \dfrac{1}{16(x-3)} + \dfrac{3}{4(x-3)^2}$

  b $\dfrac{3}{x+1} - \dfrac{3}{x} + \dfrac{3}{x^2}$

8 $A = \dfrac{1}{2}, B = -\dfrac{3}{2}$

9 $\dfrac{4}{9(x+2)} - \dfrac{4}{9(x-1)} + \dfrac{4}{3(x-1)^2}$

10 a $3 - \dfrac{13}{x+4}$ **b** $x - 2 + \dfrac{5}{x+2}$

11 $A = 1, B = 4, C = 6$

12 a $-1 - \dfrac{2}{3(x+1)} - \dfrac{1}{3(x-2)}$

  b $4x + 16 + \dfrac{64}{x-4}$

13 $A = 1, B = -6, C = 27, D = -27$

CHAPTER 3

Prior knowledge check

1 a $t = \dfrac{x}{4-k}$ **b** $t = \pm\sqrt{\dfrac{y}{3}}$

  c $t = e^{\frac{2-y}{4}}$ **d** $t = -\dfrac{1}{3}\ln\left(\dfrac{x-1}{2}\right)$

2 a $7 - 3\cos^2 x$ **b** $2\cos x\sqrt{1 - \cos^2 x}$

  c $\dfrac{\cos x}{\sqrt{1 - \cos^2 x}}$ **d** $2\cos x + 2\cos^2 x - 1$

3 a $y > 0$ **b** $0 < y \leqslant 2$

  c $-6 \leqslant y < 3$ **d** $0 < y < 1$

4 $(4,7)$ and $(-4.8, 2.6)$

Exercise 3A

1 a $y = (x+2)^2 + 1, -6 \leqslant x \leqslant 2,$ $1 \leqslant y \leqslant 17$

  b $y = (5-x)^2 - 1, x \in \mathbb{R}$ $y \geqslant -1$

  c $y = 3 - \dfrac{1}{x}, x \neq 0,$ $y \neq 3$

  d $y = \dfrac{2}{x-1}, x > 1,$ $y > 0$

  e $y = \left(\dfrac{1+2x}{x}\right)^2, x > 0,$ $y > 4$

  f $y = \dfrac{x}{1-3x}, 0 < x < \dfrac{1}{3},$ $y > 0$

2 a i $y = 20 - 10e^{\frac{1}{2}x} + e^x, x > 0$ **ii** $y > -5$

  b i $y = \dfrac{1}{e^x + 2}, x > 0$ **ii** $0 < y < \dfrac{1}{3}$

  c i $y = x^3, x > 0$ **ii** $y > 0$

3 a $y = 9x^2 - x^4,$ $0 \leqslant x \leqslant \sqrt{5},$ $0 \leqslant y \leqslant \dfrac{81}{4}$

  b

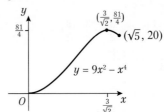

4 a i $y = \dfrac{15}{2} - \dfrac{1}{2}x$ **ii** $x > -3, y < 9$

iii

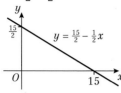

b i $y = \dfrac{1}{9}(x - 2)(x + 7)$

ii $-13 < x < 11, -\dfrac{9}{4} < y < 18$

iii

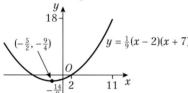

c i $y = \dfrac{1}{x - 2}$

ii $x \in \mathbb{R}, x \neq 2, y \in \mathbb{R}, y \neq 0$

iii

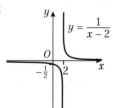

d i $y = 3x + 3$

ii $x > -1, y > 0$

iii

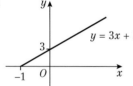

e i $y = 2 - e^x$

ii $x > 0, y < 1$

iii

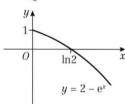

5 a $C_1: x = 1 + 2t, t = \dfrac{x - 1}{2}$

Sub t into $y = 2 + 3t$:

$y = 2 + 3\left(\dfrac{x - 1}{2}\right) = 2 + \dfrac{3}{2}x - \dfrac{3}{2} = \dfrac{3}{2}x + \dfrac{1}{2}$

$C_2: x = \dfrac{1}{2t - 3}, t = \dfrac{1 + 3x}{2x}$

Sub into $y = \dfrac{t}{2t - 3}$.

$y = \dfrac{\dfrac{1 + 3x}{2x}}{2\left(\dfrac{1 + 3x}{2x}\right) - 3} = \dfrac{\dfrac{1 + 3x}{2x}}{\dfrac{1}{x}} = \dfrac{1 + 3x}{2} = \dfrac{3}{2}x + \dfrac{1}{2}$

Therefore C_1 and C_2 represent a segment of the same straight line.

b Length of $C_1 = 3\sqrt{13}$, length of $C_2 = \dfrac{\sqrt{13}}{3}$

6 a $x \neq 2, y \leqslant -2$

b $x = \dfrac{3}{t} + 2, t = \dfrac{3}{x - 2}$

Sub into $y = 2t - 3 - t^2$

$y = 2\left(\dfrac{3}{x - 2}\right) - 3 - \left(\dfrac{3}{x - 2}\right)^2 = \dfrac{6}{x - 2} - 3 - \dfrac{9}{(x - 2)^2}$

$= \dfrac{6(x - 2) - 3(x - 2)^2 - 9}{(x - 2)^2}$

$= \dfrac{6x - 12 - 3x^2 + 12x - 12 - 9}{(x - 2)^2} = \dfrac{-3x^2 + 18x - 33}{(x - 2)^2}$

$= \dfrac{-3(x^2 - 6x + 11)}{(x - 2)^2}$ so $A = -3, b = -6, c = 11$

7 a $x = \ln(t + 3)$ $t = e^x - 3$ Sub into $y = \dfrac{1}{t + 5}$

$y = \dfrac{1}{e^x - 3 + 5} = \dfrac{1}{e^x + 2}, \quad x > 0$

b $0 < y < \dfrac{1}{3}$

8 a $y = \dfrac{x^6}{729} - \dfrac{2x^2}{9}, 0 \leqslant x \leqslant 3\sqrt{2}$

b $y = t^3 - 2t, \dfrac{dy}{dt} = 3t^2 - 2$

$0 = 3t^2 - 2 \quad t^2 = \dfrac{2}{3} \quad t = \sqrt{\dfrac{2}{3}}$

c $-\dfrac{4\sqrt{6}}{9} \leqslant f(x) \leqslant 4$

9 a $y = 4 - t^2 \Rightarrow t = \sqrt{4 - y}$

Sub into $x = t^3 - t = t(t^2 - 1)$

$x = \sqrt{4 - y}(4 - y - 1) = \sqrt{4 - y}(3 - y)$

$x^2 = (4 - y)(3 - y)^2$

$a = 4, b = 3$

b Max y is 4

Challenge

a $x^2 = \dfrac{(1 - t^2)^2}{(1 + t^2)^2}, y^2 = \dfrac{4t^2}{(1 + t^2)^2}$

$x^2 + y^2 = \dfrac{(1 - t^2)^2}{(1 + t^2)^2} + \dfrac{4t^2}{(1 + t^2)^2} = \dfrac{1 - 2t^2 + t^4}{(1 + t^2)^2} + \dfrac{4t^2}{(1 + t^2)^2}$

$= \dfrac{1 + 2t^2 + t^4}{(1 + t^2)^2} = \dfrac{(1 + t^2)^2}{(1 + t^2)^2} = 1$

So $x^2 + y^2 = 1$

b Circle, centre $(0,0)$, radius 1.

Exercise 3B

1 a $25(x + 1)^2 + 4(y - 4)^2 = 100$ **b** $y^2 = 4x^2(1 - x^2)$

c $y = 4x^2 - 2$ **d** $y = \dfrac{2x\sqrt{1 - x^2}}{1 - 2x^2}$

e $y = \dfrac{4}{x - 2}$ **f** $y^2 = 1 + \left(\dfrac{x}{3}\right)^2$

2 a $(x + 5)^2 + (y - 2)^2 = 1$

b Centre $(-5, 2)$, radius 1

c $0 \leqslant t < 2\pi$

3 Centre $(3, -1)$, radius 4

4 $(x + 2)^2 + (y - 3)^2 = 1$

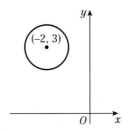

5 **a** $y = \dfrac{\sqrt{2}}{2}x + \dfrac{\sqrt{2(1 - x^2)}}{2}$

b $y = \dfrac{\sqrt{3}}{3}x - \dfrac{\sqrt{9 - x^2}}{3}$

c $y = -3x$

6 **a** $y = \dfrac{16}{x^2}$

b

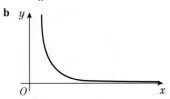

7 $y = \dfrac{9}{3 + x}$ Domain: $x > 0$

8 **a** $y = 9x(1 - 12x^2) \Rightarrow a = 9, b = 12$

b Domain: $0 < x < \dfrac{1}{3}$, Range: $-1 < y \leqslant 1$

9 $y = \sin t \cos\left(\dfrac{\pi}{6}\right) - \cos t \sin\left(\dfrac{\pi}{6}\right)$

$= \dfrac{\sqrt{3}}{2}\sin t - \dfrac{1}{2}\cos t = \dfrac{\sqrt{3\left(1 - \dfrac{x^2}{4}\right)}}{2} - \dfrac{1}{4}x$

$= \dfrac{1}{4}\left(2\sqrt{3 - \dfrac{3}{4}x^2} - x\right) = \dfrac{1}{4}\left(\sqrt{12 - 3x^2} - x\right)$

$t = 0 \Rightarrow x = 2, t = \pi \Rightarrow x = -2$, so $-2 < x < 2$.

10 **a** $y^2 = 25\left(1 - \dfrac{1}{x - 4}\right)$ **b** $x > 5, 0 < y < 5$

11 $x = -\dfrac{y}{\sqrt{9 - y^2}}, x > 0$

Challenge

$(4y^2 - 2 + 2x)^2 + 12x^2 - 3 = 0$

Exercise 3C

1

t	−5	−4	−3	−2	−1	−0.5
$x = 2t$	−10	−8	−6	−4	−2	−1
$y = \dfrac{5}{t}$	−1	−1.25	−1.67	−2.5	−5	−10

t	0.5	1	2	3	4	5
$x = 2t$	1	2	4	6	8	10
$y = \dfrac{5}{t}$	10	5	2.5	1.67	1.25	1

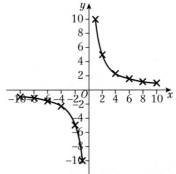

2

t	−4	−3	−2	−1	0
$x = t^2$	16	9	4	1	0
$y = \dfrac{t^3}{5}$	−12.8	−5.4	−1.6	−0.2	0

t	1	2	3	4
$x = t^2$	1	4	9	16
$y = \dfrac{t^3}{5}$	0.2	1.6	5.4	12.8

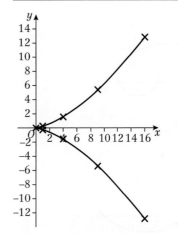

3

t	$-\dfrac{\pi}{4}$	$-\dfrac{\pi}{6}$	$-\dfrac{\pi}{12}$	0
$x = \tan t + 1$	0	0.423	0.732	1
$y = \sin t$	−0.707	−0.5	−0.259	0

t	$\dfrac{\pi}{12}$	$\dfrac{\pi}{6}$	$\dfrac{\pi}{4}$	$\dfrac{\pi}{3}$
$x = \tan t + 1$	1.268	1.577	2	2.732
$y = \sin t$	0.259	0.5	0.707	0.866

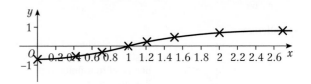

4 a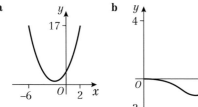

b

c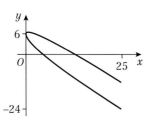

d e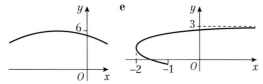

f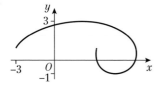

5 a $y = (3 - x)^2 - 2$

b

6 a $(x + 2)^2 + (y - 1)^2 = 81$

b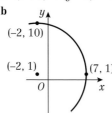

c 6π

Challenge

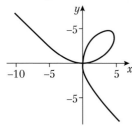

As t approaches -1 from the positive direction, the curve heads off to infinity in the 2nd quadrant.
As t approaches -1 from the negative direction, the curve heads off to infinity in the 4th quadrant.

Chapter review 3

1 a $A(4, 0), B(0, 3)$ **b** $C\left(2\sqrt{3}, \dfrac{3}{2}\right)$ **c** $\left(\dfrac{x}{4}\right)^2 + \left(\dfrac{y}{3}\right)^2 = 1$

2

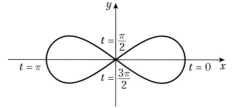

3 a $y = \ln(2\sqrt{x - 1}) - \dfrac{1}{2}, x > e^3 + 1$

b $y > 1 + \ln 2$

4 $y = -\ln(4) - 2\ln(x), 0 < x < \dfrac{1}{2}, y > 0$

5 a $(x + 3)^2 + (y - 5)^2 = 16$ **b**

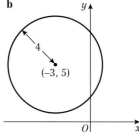

c $(0, 5 + \sqrt{7}), (0, 5 - \sqrt{7})$

6 a $x(1 + t) = 2 - 3t \Rightarrow xt + 3t = 2 - x \Rightarrow t(x + 3) = 2 - x$

$\Rightarrow t = \dfrac{2 - x}{x + 3}$

Sub into $y = \dfrac{3 + 2t}{1 + t}$

$y = \dfrac{3 + 2\left(\dfrac{2 - x}{x + 3}\right)}{1 + \left(\dfrac{2 - x}{x + 3}\right)} = \dfrac{3(x + 3) + 2(2 - x)}{x + 3 + 2 - x} = \dfrac{3x + 9 + 4 - 2x}{5}$

$= \dfrac{x + 13}{5} \Rightarrow y = \dfrac{1}{5}x + \dfrac{13}{5}$

This is in the form $y = mx + c$, therefore the curve C is a straight line.

b $\dfrac{4\sqrt{26}}{5}$

7 a $y = 2\sqrt{x + 2}$

b Domain: $-2 \le x \le 2$, Range: $0 \le y \le 4$

c

8 a $\cos t = \dfrac{x}{2}, \sin t = \dfrac{y + 5}{2}$

$\left(\dfrac{x}{2}\right)^2 + \left(\dfrac{y + 5}{2}\right)^2 = 1 \Rightarrow x^2 + (y + 5)^2 = 4$

Since $0 \le t \le \pi$, the curve C forms half of a circle.

Online Worked solutions are available in SolutionBank.

b

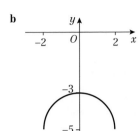

c 2π

9 a $y = x^3 + 4x^2 + 4x$ **b**

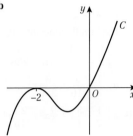

CHAPTER 4

Prior knowledge check

1 a $1 + 35x + 525x^2 + 4375x^3$
b $9\,765\,625 - 39\,062\,500x + 70\,312\,500x^2$
$- 75\,000\,000x^3$
c $64 + 128x + 48x^2 - 80x^3$

2 a $\dfrac{4}{1 + 2x} + \dfrac{3}{1 - 5x}$ **b** $\dfrac{12}{1 + 2x} - \dfrac{13}{(1 + 2x)^2}$

c $\dfrac{8}{3x - 4} + \dfrac{56}{(3x - 4)^2}$

Exercise 4A

1 a i $1 - 4x + 10x^2 - 20x^3\ldots$ **ii** $|x| < 1$
b i $1 - 6x + 21x^2 - 56x^3\ldots$ **ii** $|x| < 1$
c i $1 + \dfrac{x}{2} - \dfrac{x^2}{8} + \dfrac{x^3}{16}\ldots$ **ii** $|x| < 1$
d i $1 + \dfrac{5x}{3} + \dfrac{5x^2}{9} - \dfrac{5x^3}{81}\ldots$ **ii** $|x| < 1$
e i $1 - \dfrac{x}{4} + \dfrac{5x^2}{32} - \dfrac{15x^3}{128}\ldots$ **ii** $|x| < 1$
f i $1 - \dfrac{3x}{2} + \dfrac{15x^2}{8} - \dfrac{35x^3}{16}\ldots$ **ii** $|x| < 1$

2 a i $1 - 9x + 54x^2 - 270x^3\ldots$ **ii** $|x| < \dfrac{1}{3}$
b i $1 - \dfrac{5x}{2} + \dfrac{15x^2}{4} - \dfrac{35x^3}{8}\ldots$ **ii** $|x| < 2$
c i $1 + \dfrac{3x}{2} - \dfrac{3x^2}{8} + \dfrac{5x^3}{16}\ldots$ **ii** $|x| < \dfrac{1}{2}$
d i $1 - \dfrac{35x}{3} + \dfrac{350x^2}{9} - \dfrac{1750x^3}{81}\ldots$ **ii** $|x| < \dfrac{1}{5}$
e i $1 - 4x + 20x^2 - \dfrac{320x^3}{3}\ldots$ **ii** $|x| < \dfrac{1}{6}$
f i $1 + \dfrac{5x}{4} + \dfrac{5x^2}{4} + \dfrac{55x^3}{48}\ldots$ **ii** $|x| < \dfrac{4}{3}$

3 a i $1 - 2x + 3x^2 - 4x^3\ldots$ **ii** $|x| < 1$
b i $1 - 12x + 90x^2 - 540x^3\ldots$ **ii** $|x| < \dfrac{1}{3}$
c i $1 - \dfrac{x}{2} - \dfrac{x^2}{8} - \dfrac{x^3}{16}\ldots$ **ii** $|x| < 1$
d i $1 - x - x^2 - \dfrac{5x^3}{3}\ldots$ **ii** $|x| < \dfrac{1}{3}$

e i $1 - \dfrac{x}{4} + \dfrac{3x^2}{32} - \dfrac{5x^3}{128}\ldots$ **ii** $|x| < 2$
f i $1 + \dfrac{4x}{3} + \dfrac{20x^2}{9} + \dfrac{320x^3}{81}\ldots$ **ii** $|x| < \dfrac{1}{2}$

4 a Expansion of $(1 - 2x)^{-1} = 1 + 2x + 4x^2 + 8x^3 + \ldots$
Multiply by $(1 + x) = 1 + 3x + 6x^2 + 12x^3 + \ldots$
b $|x| < \dfrac{1}{2}$

5 a $1 + \dfrac{3}{2}x - \dfrac{9}{8}x^2 + \dfrac{27}{16}x^3$
b $f(x) = \sqrt{\dfrac{103}{100}} = \dfrac{\sqrt{103}}{\sqrt{100}} = \dfrac{\sqrt{103}}{10}$
c $3.1 \times 10^{-6}\%$

6 a $a = \pm 8$ **b** $\pm 160x^3$

7 For small values of x ignore powers of x^3 and higher.
$(1 + x)^{\frac{1}{2}} = 1 + \dfrac{x}{2} - \dfrac{x^2}{8} + \ldots$, $(1 - x)^{-\frac{1}{2}} = 1 + \dfrac{x}{2} + \dfrac{3x^2}{8} + \ldots$
$\sqrt{\dfrac{1 + x}{1 - x}} = 1 + \dfrac{x}{2} - \dfrac{x^2}{8} + \dfrac{x}{2} + \dfrac{x^2}{4} + \dfrac{3x^2}{8} + \ldots = 1 + x + \dfrac{x^2}{2}$

8 a $2 - 42x + 114x^2$
b 0.052%
c The expansion is only valid for $|x| < \dfrac{1}{5}$. $|0.5|$ is not less than $\dfrac{1}{5}$.

9 a $1 - \dfrac{9}{2}x + \dfrac{27}{8}x^2 + \dfrac{27}{16}x^3$
b $0.97^{\frac{3}{2}} = \left(\dfrac{\sqrt{97}}{10}\right)^3 = \dfrac{97\sqrt{97}}{1000}$
c 9.84886

Challenge

a $1 - \dfrac{1}{2x} + \dfrac{3}{8x^2}$
b $h(x) = \left(\dfrac{10}{9}\right)^{-\frac{1}{2}} = \left(\dfrac{9}{10}\right)^{\frac{1}{2}} = \dfrac{3}{\sqrt{10}} = \dfrac{3\sqrt{10}}{10}$
c 3.16

Exercise 4B

1 a i $2 + \dfrac{x}{2} - \dfrac{x^2}{16} + \dfrac{x^3}{64}$ **ii** $|x| < 2$
b i $\dfrac{1}{2} - \dfrac{x}{4} + \dfrac{x^2}{8} - \dfrac{x^3}{16}$ **ii** $|x| < 2$
c i $\dfrac{1}{16} + \dfrac{x}{32} - \dfrac{3x^2}{256} + \dfrac{x^3}{256}$ **ii** $|x| < 4$
d i $3 + \dfrac{x}{6} - \dfrac{x^2}{216} + \dfrac{x^3}{3888}$ **ii** $|x| < 9$
e i $\dfrac{\sqrt{2}}{2} - \dfrac{\sqrt{2}}{8}x + \dfrac{3\sqrt{2}}{64}x^2 - \dfrac{5\sqrt{2}}{256}x^3$ **ii** $|x| < 2$
f $\dfrac{5}{3} - \dfrac{10}{9}x + \dfrac{20}{27}x^2 - \dfrac{40}{81}x^3$ **ii** $|x| < \dfrac{3}{2}$
g i $\dfrac{1}{2} + \dfrac{1}{4}x - \dfrac{1}{8}x^2 + \dfrac{1}{16}x^3$ **ii** $|x| < 2$
h i $\sqrt{2} + \dfrac{3\sqrt{2}}{4}x + \dfrac{15\sqrt{2}}{32}x^2 + \dfrac{51\sqrt{2}}{128}x^3$ **ii** $|x| < 1$

2 $\dfrac{1}{25} - \dfrac{8}{125}x + \dfrac{48}{625}x^2 - \dfrac{256}{3125}x^3$

3 a $2 - \dfrac{x}{4} - \dfrac{x^2}{64}$
b $m(x) = \sqrt{\dfrac{35}{9}} = \dfrac{\sqrt{35}}{\sqrt{9}} = \dfrac{\sqrt{35}}{3}$

c 5.91609 (correct to 5 decimal places),
% error = 1.38×10^{-4}%

4 a $a = \dfrac{1}{9}, b = -\dfrac{2}{81}$ **b** $\dfrac{5}{486}$

5 For small values of x ignore powers of x^3 and higher.

$(4 - x)^{-1} = \dfrac{1}{4} + \dfrac{x}{16} + \dfrac{x^2}{64} + \dots$

Multiply by $(3 + 2x - x^2) = \dfrac{3}{4} + \dfrac{x}{2} - \dfrac{x^2}{4} + \dfrac{3x}{16} + \dfrac{x^2}{8} + \dfrac{3x^2}{64}$

$= \dfrac{3}{4} + \dfrac{11}{16}x - \dfrac{5}{64}x^2$

6 a $\dfrac{1}{\sqrt{5}} - \dfrac{x}{5\sqrt{5}} + \dfrac{3x^2}{50\sqrt{5}}$ **b** $-\dfrac{1}{\sqrt{5}} + \dfrac{11x}{5\sqrt{5}} - \dfrac{23x^2}{50\sqrt{5}}$

7 a $2 - \dfrac{3}{32}x - \dfrac{27}{4096}x^2$ **b** 1.991

8 a $\dfrac{3}{4 - 2x} = \dfrac{3}{4} + \dfrac{3x}{8} + \dfrac{3x^2}{16}, \dfrac{2}{3 + 5x} = \dfrac{2}{3} - \dfrac{10x}{9} + \dfrac{50x^2}{27}$

$\dfrac{3}{4 - 2x} - \dfrac{2}{3 + 5x} = \dfrac{1}{12} + \dfrac{107}{72}x - \dfrac{719}{432}x^2$

b 0.0980311
c 0.0032%

Exercise 4C

1 a $\dfrac{4}{1 - x} - \dfrac{4}{2 + x}$ **b** $2 + 5x + \dfrac{7}{2}x^2$

c valid $|x| < 1$

2 a $-\dfrac{2}{2 + x} + \dfrac{4}{(2 + x)^2}$ **b** $B = \dfrac{1}{2}, C = -\dfrac{3}{8}$

c $|x| < 2$

3 a $\dfrac{2}{1 + x} + \dfrac{3}{1 - x} - \dfrac{4}{2 + x}$ **b** $3 + 2x + \dfrac{9}{2}x^2 + \dfrac{5}{4}x^3$

c $|x| < 1$

4 a $A = -\dfrac{14}{5}$ and $B = \dfrac{9}{5}$ **b** $-1 + 11x + 5x^2$

5 a $2 - \dfrac{1}{x + 5} + \dfrac{6}{x - 4}$ **b** $\dfrac{3}{10} - \dfrac{67}{200}x - \dfrac{407}{4000}x^2$

c $|x| < 4$

6 a $A = 3, B = -2$ and $C = 3$ **b** $\dfrac{5}{6} - \dfrac{19}{36}x - \dfrac{97}{216}x^2$

7 a $A = -\dfrac{7}{9}, B = \dfrac{28}{3}$ and $C = \dfrac{8}{9}$ **b** $11 + 38x + 116x^2$

c 0.33%

Chapter review 4

1 a i $1 - 12x + 48x^2 - 64x^3$ **ii** all x

b i $4 + \dfrac{x}{8} - \dfrac{x^2}{512} + \dfrac{x^3}{16384}$ **ii** $|x| < 16$

c i $1 + 2x + 4x^2 + 8x^3$ **ii** $|x| < \dfrac{1}{2}$

d i $2 - 3x + \dfrac{9x^2}{2} - \dfrac{27x^3}{4}$ **ii** $|x| < \dfrac{2}{3}$

e i $2 + \dfrac{x}{4} + \dfrac{3x^2}{64} + \dfrac{5x^3}{512}$ **ii** $|x| < 4$

f i $1 - 2x + 6x^2 - 18x^3$ **ii** $|x| < \dfrac{1}{3}$

g i $1 + 4x + 8x^2 + 12x^3$ **ii** $|x| < 1$

h i $-3 - 8x - 18x^2 - 38x^3$ **ii** $|x| < \dfrac{1}{2}$

2 $1 - \dfrac{x}{4} - \dfrac{x^2}{32} - \dfrac{x^3}{128}$

3 a $1 + \dfrac{x}{2} - \dfrac{x^2}{8} + \dfrac{x^3}{16}$ **b** $\dfrac{1145}{512}$

4 a $c = -9, d = 36$ **b** 1.282

c calculator = 1.28108713, approximation is correct to 2 decimal places.

5 a $a = 4$ or $a = -4$

b coefficient of $x^3 = 4$, coefficient of $x^3 = -4$

6 a $1 - 3x + 9x^2 - 27x^3$

b $(1 + x)(1 - 3x + 9x^2 - 27x^3)$
$= 1 - 3x + 9x^2 - 27x^3 + x - 3x^2 + 9x^3$
$= 1 - 2x + 6x^2 - 18x^3$

c $x = 0.01, 0.980\,58$

7 a $n = -2, a = 3$ **b** -108

c $|x| < \dfrac{1}{3}$

8 For small values of x ignore powers of x^3 and higher.

$\dfrac{1}{\sqrt{4 + x}} = \dfrac{1}{2} - \dfrac{x}{16} + \dfrac{3x^2}{256}, \dfrac{3}{\sqrt{4 + x}} = \dfrac{3}{2} - \dfrac{3}{16}x + \dfrac{9}{256}x^2$

9 a $\dfrac{1}{2} + \dfrac{x}{16} + \dfrac{3}{256}x^2$ **b** $\dfrac{1}{2} + \dfrac{17}{16}x + \dfrac{35}{256}x^2$

10 a $\dfrac{1}{2} - \dfrac{3}{4}x + \dfrac{9}{8}x^2 - \dfrac{27}{16}x^3$ **b** $\dfrac{1}{2} - \dfrac{x}{4} + \dfrac{3}{8}x^2 - \dfrac{9}{16}x^3$

11 a $\dfrac{1}{2} - \dfrac{x}{16} + \dfrac{3x^2}{256} - \dfrac{5x^3}{2048}$ **b** 0.6914

12 $\dfrac{1}{27} - \dfrac{4}{27}x + \dfrac{32}{81}x^2$

13 a $A = 1, B = -4, C = 3$ **b** $-\dfrac{3}{8} - \dfrac{51}{64}x + \dfrac{477}{512}x^2$

14 a $A = 3$ and $B = 2$ **b** $5 - 28x + 144x^2$

15 a $10 - 2x + \dfrac{5}{2}x^2 - \dfrac{11}{4}x^3$, so $B = \dfrac{5}{2}$ and C $= -\dfrac{11}{4}$

b Percent error = 0.0027%

Challenge

$1 - \dfrac{3x^2}{2} + \dfrac{27x^4}{8} - \dfrac{135x^6}{16}$

Review exercise 1

1 Assumption: there are finitely many prime numbers, p_1, p_2, p_3 up to p_n. Let $X = (p_1 \times p_2 \times p_3 \times \dots \times p_n) + 1$
None of the prime numbers $p_1, p_2, \dots p_n$ can be a factor of X as they all leave a remainder of 1 when X is divided by them. But X must have at least one prime factor. This is a contradiction.
So there are infinitely many prime numbers.

2 Assumption: $x = \dfrac{a}{b}$ is a solution to the equation, where a and b are integers with no common factors.
$\left(\dfrac{a}{b}\right)^2 - 2 = 0 \Rightarrow \dfrac{a^2}{b^2} = 2 \Rightarrow a^2 = 2b^2$
So a^2 is even, which implies that a is even.
Write $a = 2n$ for some integer n.
$(2n)^2 = 2b^2 \Rightarrow 4n^2 = 2b^2 \Rightarrow 2n^2 = b^2$
So b^2 is even, which implies that b is even.
This contradicts the assumption that a and b have no common factor.
Hence there are no rational solutions to the equation.

3 Assumption: if n is odd then $3n^2 + 2$ is even.
Let $n = 2k + 1$ where k is an integer.
Then $3(2k + 1)^2 + 2 = 3(4k^2 + 4k + 1) + 2 = 12(k^2 + k) + 5$
This is an even number plus an odd number which must be odd. The contradicts the assumption made.
Therefore if n is odd then $3n^2 + 2$ is odd.

Online Worked solutions are available in SolutionBank.

4 Assumption: $\sqrt{5}$ is rational.

Let $\sqrt{5} = \dfrac{a}{b}$ for integers a and b. Also assume that this fraction is in its simplified form and there are no common factors.

Then $5 = \dfrac{a^2}{b^2}$ or $a^2 = 5b^2$. Hence a^2 must be a multiple of 5. Since a is an integer it follows that a is also a multiple of 5. So let $a = 5c$ where c is an integer.

Then $a^2 = 25c^2$, and so $5b^2 = 25c^2$ which leads to $b^2 = 5c^2$

Now b^2 is a multiple of 5, and so b is a multiple of 5.

If a and b are both multiples of 5 then this contradicts the initial statement of there being no common factors.

Hence $\sqrt{5}$ is irrational.

5 $\dfrac{-1}{x-1} + \dfrac{4}{2x-3}$

6 $P = 2, Q = -1, R = -1$

7 $A = \dfrac{2}{9}, B = \dfrac{2}{9}, C = \dfrac{2}{3}$

8 $A = 3, B = 1, C = -2$

9 $d = 3, e = 6, f = -14$

10 $p(x) = 6 - \dfrac{2}{1-x} + \dfrac{5}{1+2x}$

11 $\dfrac{13}{2(x+3)} - \dfrac{5}{2(x+1)}$

12 $A = 4, B = 27, C = -7, D = -2$

13 a $\dfrac{5x+3}{(2x-3)(x-2)} \equiv \dfrac{3}{2x-3} + \dfrac{1}{x+2}$ **b** $\ln 54$

14 a $x \neq 1, y \geq -1.25$

b $t = \dfrac{-4}{x-1} = \dfrac{4}{1-x}$

$y = \left(\dfrac{4}{1-x}\right)^2 - 3\left(\dfrac{4}{1-x}\right) + 1$

$y = \dfrac{16}{(1-x)^2} - \dfrac{12(1-x)}{(1-x)^2} + \dfrac{(1-x)^2}{(1-x)^2}$

$y = \dfrac{16 - 12 + 12x + 1 - 2x + x^2}{(1-x)^2}$

$y = \dfrac{x^2 + 10x + 5}{(1-x)^2} \Rightarrow a = 1, b = 10, c = 5$

15 a $t = e^x - 2$

$y = \dfrac{3t}{t+3} = \dfrac{3e^x - 6}{e^x + 1}$

$t > 4 \Rightarrow e^x - 2 > 4 \Rightarrow e^x > 6 \Rightarrow x > \ln 6$

b $t = 4 \Rightarrow y = \dfrac{12}{7}, x \to \infty, y \to 3, \dfrac{12}{7} < y < 3$

16 $x = \dfrac{1}{1+t} \Rightarrow t = \dfrac{1-x}{x}$

$y = \dfrac{1}{1 - \dfrac{1-x}{x}} = \dfrac{x}{x - (1-x)} = \dfrac{x}{2x-1}$

17 a $y = \cos 3t = \cos(2t + t) = \cos 2t \cos t - \sin 2t \sin t$

$= (2\cos^2 t - 1)\cos t - 2\sin^2 t \cos t$

$= 2\cos^3 t - \cos t - 2(1 - \cos^2 t)\cos t$

$= 4\cos^3 t - 3\cos t$

$x = 2\cos t \Rightarrow \cos t = \dfrac{x}{2}$

$y = 4\left(\dfrac{x}{2}\right)^3 - 3\left(\dfrac{x}{2}\right) = \dfrac{x}{2}(x^2 - 3)$

b $0 \leq x \leq 2, -1 \leq y \leq 1$

18 a $y = \sin\left(t + \dfrac{\pi}{6}\right) = \sin t \cos\dfrac{\pi}{6} + \cos t \sin\dfrac{\pi}{6}$

$= \dfrac{\sqrt{3}}{2}\sin t + \dfrac{1}{2}\cos t$

$= \dfrac{\sqrt{3}}{2}\sin t + \dfrac{1}{2}\sqrt{1 - \sin^2 t}$

$= \dfrac{\sqrt{3}}{2}x + \dfrac{1}{2}\sqrt{1 - x^2}$

$-1 < \sin t < 1 \Rightarrow -1 < x < 1$

b $A = (-0.5, 0), B = (0, 0.5)$

19 a $y = 2\left(\dfrac{x}{3}\right)^2 - 1, -3 \leq x \leq 3$

b

$y = 2\left(\dfrac{x}{3}\right)^2 - 1$

20 $\tan\left(x + \dfrac{\pi}{6}\right) = \dfrac{1}{6} \Rightarrow \dfrac{\tan x + \dfrac{\sqrt{3}}{3}}{1 - \dfrac{\sqrt{3}}{3}\tan x} = \dfrac{1}{6}$

$6\tan x + 2\sqrt{3} = 1 - \dfrac{\sqrt{3}}{3}\tan x$

$\left(\dfrac{18 + \sqrt{3}}{3}\right)\tan x = 1 - 2\sqrt{3}$

$\tan x = \dfrac{3 - 6\sqrt{3}}{18 + \sqrt{3}} \times \dfrac{18 - \sqrt{3}}{18 - \sqrt{3}} = \dfrac{72 - 111\sqrt{3}}{321}$

21 a $\sin(x + 30°) = 2\sin(x - 60°)$

$\sin x \cos 30° + \cos x \sin 30°$

$= 2(\sin x \cos 60° - \cos x \sin 60°)$

$\dfrac{\sqrt{3}}{2}\sin x + \dfrac{1}{2}\cos x = 2\left(\dfrac{1}{2}\sin x - \dfrac{\sqrt{3}}{2}\cos x\right)$

$\sqrt{3}\sin x + \cos x = 2\sin x - 2\sqrt{3}\cos x$

$(-2 + \sqrt{3})\sin x = (-1 - 2\sqrt{3})\cos x$

$\dfrac{\sin x}{\cos x} = \dfrac{-1 - 2\sqrt{3}}{-2 + \sqrt{3}} = \dfrac{-1 - 2\sqrt{3}}{-2 + \sqrt{3}} \times \dfrac{-2 - \sqrt{3}}{-2 - \sqrt{3}}$

$= \dfrac{2 + 4\sqrt{3} + \sqrt{3} + 6}{4 + 2\sqrt{3} - 2\sqrt{3} - 3} = 8 + 5\sqrt{3}$

b $8 - 5\sqrt{3}$

22 a $\sin 165° = \sin(120° + 45°)$

$= \sin 120° \cos 45° + \cos 120° \sin 45°$

$= \dfrac{\sqrt{3}}{2} \times \dfrac{1}{\sqrt{2}} + \dfrac{-1}{2} \times \dfrac{1}{\sqrt{2}} = \dfrac{\sqrt{3} - 1}{2\sqrt{2}} = \dfrac{\sqrt{6} - \sqrt{2}}{4}$

b $\text{cosec } 165° = \dfrac{1}{\sin 165°}$

$= \dfrac{4}{(\sqrt{6} - \sqrt{2})} \times \dfrac{(\sqrt{6} + \sqrt{2})}{(\sqrt{6} + \sqrt{2})} = \dfrac{4(\sqrt{6} + \sqrt{2})}{6 - 2} = \sqrt{6} + \sqrt{2}$

23 a $(1-x)^{-\frac{1}{2}} = 1 + \left(-\dfrac{1}{2}\right)(-x) + \dfrac{\left(-\dfrac{1}{2}\right)\left(-\dfrac{3}{2}\right)}{2!}(-x)^2$

$+ \dfrac{\left(-\dfrac{1}{2}\right)\left(-\dfrac{3}{2}\right)\left(-\dfrac{5}{2}\right)}{3!}(-x)^3 + \ldots$

$= 1 + \dfrac{1}{2}x + \dfrac{3}{8}x^2 + \dfrac{5}{16}x^3 + \ldots$

b $|x| < 1$ Accept $-1 < x < 1$

24 a $a = 9$ $n = -\dfrac{36}{54} = -\dfrac{2}{3}$

b -360

c $-\dfrac{1}{9} < x < \dfrac{1}{9}$

25 a $1 + 6x + 6x^2 - 4x^3$

b $\left(1 + 4\left(\dfrac{3}{100}\right)\right)^{\frac{3}{2}} = \left(\dfrac{112}{100}\right)^{\frac{3}{2}} = \left(\sqrt{\dfrac{112}{100}}\right)^3 = \left(\dfrac{\sqrt{112}}{10}\right)^3$

$= \dfrac{112\sqrt{112}}{1000}$

c 10.58296 **d** 0.00039%

26 $\dfrac{1}{27} - \dfrac{x}{27} + \dfrac{2x^2}{81} - \dfrac{8x^3}{729}$

27 a $(4 - 9x)^{\frac{1}{2}} = 2\left(1 - \dfrac{9}{4}x\right)^{\frac{1}{2}} = 2 - \dfrac{9}{4}x - \dfrac{81}{64}x^2$

b $\sqrt{4 - 9\left(\dfrac{1}{100}\right)} = \sqrt{\dfrac{391}{100}} = \dfrac{\sqrt{391}}{10}$

c Approximate: 1.97737 correct to 5 decimal places.

28 a $a = 2, b = -1, c = \dfrac{3}{16}$ **b** $\dfrac{1}{8}$

$a = -2, b = 1, c = \dfrac{3}{16}$

29 a $A = 1, B = 2$ **b** $3 - x + 11x^2 - \dots$

30 a $A = -\dfrac{3}{2}, B = \dfrac{1}{2}$ **b** $-1 - x + 4x^3 + \dots$

31 a $A = 2, B = 10, C = 1$ **b** $\dfrac{25}{9} - \dfrac{25}{27}x + \dfrac{25}{9}x^2 + \dots$

32 a $A = 2, B = 5, C = -2$

b $2 + 5(4 + x)^{-1} - 2(3 + 2x)^{-1}$

$= 2 + \dfrac{5}{4}\left(1 + \dfrac{x}{4}\right)^{-1} - \dfrac{2}{3}\left(1 + \dfrac{2}{3}x\right)^{-1}$

$= 2 + \dfrac{5}{4}\left(1 - \dfrac{x}{4} + \dfrac{x^2}{16}\right) - \dfrac{2}{3}\left(1 - \dfrac{2}{3}x + \dfrac{4}{9}x^2\right)$

$= \dfrac{31}{12} + \dfrac{19}{144}x - \dfrac{377}{1728}x^2$

Challenge

1 Assumption: There exists $a, b \in \mathbb{Z}$ such that $a^2 - 8b = 2$
First let $a^2 = 2 + 8b = 2(4b + 1)$ which means that a^2 is even.

Since a^2 is even, it follows that must also be even.

Then let $a = 2c$ where c is an integer.

So $(2c)^2 = 2(4b + 1)$, or $2c^2 - 4b = 1$

Since we now have $2(c^2 - 2b) = 1$, and that also $b, c \in \mathbb{Z}$ it follows that from our statement that 1 must be even.

Since we know 1 isn't even, this contradicts our assumption, and so $a^2 - 8b \neq 2$

2 $2x^2 + 2 - \dfrac{5}{2(x + 1)} + \dfrac{5}{2(x - 1)}$

3 a $y = 3x - 4x^3$

b $0 \leqslant x \leqslant 1, -1 \leqslant y \leqslant 1$

4 $\sqrt{98} \approx 9.899495$ and $\cos\dfrac{\pi}{4} \approx 0.707107$

CHAPTER 5

Prior knowledge check

1 a $2\cos 2x$ **b** $\dfrac{2x(1 - 3x)^3}{(1 - 9x)}$ **c** $2e^{2x + 3}$

2 $\dfrac{3}{4}$ **3** $(0, 2), \left(0, \dfrac{179}{27}\right), (11.1, 0)$

4 $0.588, 3.73$

Exercise 5A

1 a $\dfrac{2t - 3}{2}$ **b** $\dfrac{6t^2}{6t} = t$ **c** $\dfrac{4}{1 + 6t}$ **d** $\dfrac{15t^3}{2}$

e $-3t^3$ **f** $t(1 - t)$ **g** $\dfrac{2t}{t^2 - 1}$ **h** $\dfrac{2}{(t^2 + 2t)e^t}$

i $-\dfrac{3}{4}\tan 3t$ **j** $4\tan t$ **k** $\operatorname{cosec} t$ **l** $\cot t$

m $\dfrac{1}{te^t}$ **n** $2t^2$ **o** $\dfrac{1}{e^t}$

2 a $y = -\dfrac{1}{2}x + \dfrac{3}{2} - \pi$ **b** $2y + 5x = 57$

3 a $x = 1$ **b** $y + \sqrt{3}x = \sqrt{3}$

4 $(0, 0)$ and $(-2, -4)$

5 a $y = \dfrac{1}{4}x$

b $\dfrac{dy}{dx} = \dfrac{1}{2}e^{-t} = 0 \Rightarrow e^{-t} = 0$

No solution, therefore no stationary points.

6 $y = x + 7$

7 a $-\dfrac{1}{2}\sec t \operatorname{cosec}^3 t$ **b** $8x + \sqrt{3}y - 10 = 0$

8 a $\dfrac{\pi}{3}$

b $\dfrac{dy}{dt} = -4\cot 2t \operatorname{cosec} 2t, \dfrac{dx}{dt} = 4\cos t$

$\dfrac{dy}{dx} = \dfrac{-4\cot 2t \operatorname{cosec} 2t}{4\cos t} = \dfrac{-\cot 2t \operatorname{cosec} 2t}{\cos t}$

At $t = \dfrac{\pi}{3}, \dfrac{dy}{dx} = \dfrac{4}{3}$

Gradient of normal: $-\dfrac{3}{4}$

Equation of normal:

$y - \dfrac{4\sqrt{3}}{3} = -\dfrac{3}{4}(x - 2\sqrt{3}) \Rightarrow 9x + 12y - 34\sqrt{3} = 0$

9 a $(30, 101)$ **b** $y = 2x + 41$

c $t^2 - 10t + 5 = 2(t^2 + t) + 41$

$t^2 - 10t + 5 = 2t^2 + 2t + 41$

$0 = t^2 + 12t + 36$

Discriminant $= 12^2 - 4 \times 1 \times 36 = 144 - 144 = 0$

Therefore the curve and the line only intersects once.
Therefore it does not intersect the curve again.

10 a $-2\sqrt{2}\sin t$ **b** $x - \sqrt{6}y - 2\sqrt{3} = 0$

c $2\sin t - \sqrt{12}\cos 2t - 2\sqrt{3} = 0$

$\sin t - \sqrt{3}\cos 2t - \sqrt{3} = 0$

$2\sqrt{3}\sin^2 t + \sin t - 2\sqrt{3} = 0$

$(2\sin t - \sqrt{3})(\sqrt{3}\sin t + 2) = 0$

$\sin t = \dfrac{\sqrt{3}}{2}\left(\sin t \neq -\dfrac{2}{\sqrt{3}}\right) \Rightarrow t = \dfrac{\pi}{3}$ or $\dfrac{2\pi}{3}$

B is when $t = \dfrac{2\pi}{3}: \left(2\sin\dfrac{2\pi}{3}, \sqrt{2}\cos\dfrac{4\pi}{3}\right) = \left(\sqrt{3}, -\dfrac{1}{\sqrt{2}}\right)$

Same point as A, so l only intersects C once.

11 a $-\dfrac{\cos 2t}{\sin t}$ **b** $y = -x + \dfrac{3\sqrt{3}}{4}$

c $y = -x$ and $y = -x - \dfrac{3\sqrt{3}}{4}$

Exercise 5B

1 Letting $u = y^n$, $\dfrac{du}{dy} = ny^{n-1}$

$\dfrac{d}{dx}(y^n) = \dfrac{du}{dx} = \dfrac{du}{dy} \times \dfrac{dy}{dx} = ny^{n-1}\dfrac{dy}{dx}$

2 $\dfrac{d}{dx}(xy) = x\dfrac{d}{dx}(y) + \dfrac{d}{dx}(x)y = x\dfrac{dy}{dx} + 1 \times y = x\dfrac{dy}{dx} + y$

3 a $-\dfrac{2x}{3y^2}$ **b** $-\dfrac{x}{5y}$ **c** $\dfrac{-3-x}{5y-4}$

d $\dfrac{4-6xy}{3x^2+3y^2}$ **e** $\dfrac{3x^2-2y}{6y-2+2x}$ **f** $\dfrac{3x^2-y}{2+x}$

g $\dfrac{4(x-y)^3-1}{1+4(x-y)^3}$ **h** $\dfrac{e^x y - e^y}{xe^y - e^x}$ **i** $\dfrac{-2\sqrt{xy}-y}{4y\sqrt{xy}+x}$

4 $y = -\dfrac{7}{9}x + \dfrac{23}{9}$ **5** $y = 2x - 2$

6 $(3, 1)$ and $(3, 3)$ **7** $3x + 2y + 1 = 0$

8 $2 - 3\ln 3$ **9** $\dfrac{1}{4}(4 + 3\ln 3)$

10 a $\dfrac{\cos x}{\sin y}$ **b** $\left(\dfrac{\pi}{2}, \dfrac{2\pi}{3}\right)$ and $\left(\dfrac{\pi}{2}, -\dfrac{2\pi}{3}\right)$

11 a $\dfrac{3+3ye^{-3x}}{e^{-3x}-2y}$

b At O, $\dfrac{dy}{dx} = \dfrac{3-0}{e^0-0} = 3$

So the tangent is $y - 0 = 3(x - 0)$, or $y = 3x$.

Challenge

a $6 + 2y\dfrac{dy}{dx} + 2y + 2x\dfrac{dy}{dx} = 2x \Rightarrow \dfrac{dy}{dx} = \dfrac{x-y-3}{y+x}$

So $\dfrac{dy}{dx} = 0 \Leftrightarrow x - y = 3$

Substitute: $6x + (x-3)^2 + 2x(x-3) = x^2$

So $2x^2 - 6x + 9 = 0$

Discriminant $= -36$, so no real solutions to quadratic.

Therefore no points on C s.t. $\dfrac{dy}{dx} = 0$

b $(0, 0)$ and $(3, -3)$

Exercise 5C

1 6π **2** $15e^2$ **3** $-\dfrac{9}{2}$ **4** $\dfrac{8}{9\pi}$ **5** $\dfrac{dP}{dt} = kP$

6 $\dfrac{dy}{dx} = kxy$; at $(4, 2)$ $\dfrac{dy}{dx} = \dfrac{1}{2}$, so $8k = \dfrac{1}{2}$, $k = \dfrac{1}{16}$

Therefore $\dfrac{dy}{dx} = \dfrac{xy}{16}$

7 $\dfrac{dV}{dt} =$ rate in $-$ rate out $= 30 - \dfrac{2}{15}V \Rightarrow 15\dfrac{dV}{dt} = 450 - 2V$

So $-15\dfrac{dV}{dt} = 2V - 450$

8 $\dfrac{dQ}{dt} = -kQ$ **9** $\dfrac{dx}{dt} = \dfrac{k}{x^2}$

10 a Circumference, C, $= 2\pi r$, so $\dfrac{dC}{dt} = 2\pi \times 0.4$

$= 0.8\pi\,\text{cm s}^{-1}$

Rate of increase of circumference with respect to time.

b $8\pi\,\text{cm}^2\,\text{s}^{-1}$ **c** $\dfrac{25}{\pi}\,\text{cm}$

11 a $0.070\,\text{cm per second}$ **b** $20.5\,\text{cm}^3$

12 $\dfrac{dV}{dt} \propto \sqrt{V} \Rightarrow \dfrac{dV}{dt} = -k_1\sqrt{V}$, $V \propto h \Rightarrow h = k_2 V$

$\dfrac{dh}{dt} = \dfrac{dh}{dV} \times \dfrac{dV}{dt} = k_2 \times (-k_1\sqrt{V}) = -k_1 k_2\sqrt{\dfrac{h}{k_2}}$

$= \dfrac{-k_1 k_2}{\sqrt{k_2}}\sqrt{h} = -k\sqrt{h}$

13 a $V = \left(\dfrac{A}{6}\right)^{\frac{3}{2}}$ **b** $\dfrac{1}{4}\left(\dfrac{A}{6}\right)^{\frac{1}{2}}$

c $\dfrac{dV}{dt} = \dfrac{dV}{dA} \times \dfrac{dA}{dt} = \dfrac{1}{4}\left(\dfrac{A}{6}\right)^{\frac{1}{2}} \times 2 = \dfrac{1}{2}\left(V^{\frac{2}{3}}\right)^{\frac{1}{2}} = \dfrac{1}{2}V^{\frac{1}{3}}$

14 $V = \dfrac{\pi}{3}r^2 h = \dfrac{\pi}{3}(h\tan 30°)^2 h = \dfrac{\pi}{9}h^3$

$\dfrac{dh}{dt} = \dfrac{dh}{dV} \times \dfrac{dV}{dt} = \dfrac{1}{\frac{dh}{dV}} \times \dfrac{dV}{dt} = \dfrac{1}{\frac{\pi}{3}h^2} \times (-6) = -\dfrac{18}{\pi h^2}$

So $\dfrac{dh}{dt} \propto \dfrac{1}{h^2}$

Chapter review 5

1 a $\dfrac{dy}{dx} = -\dfrac{4}{t^3}$ **b** $y = 2x - 8$

2 $3y + x = 33$ **3** $y = \dfrac{2}{3}x + \dfrac{1}{3}$

4 a $\dfrac{dx}{dt} = -2\sin t + 2\cos 2t$; $\dfrac{dy}{dt} = -\sin t - 4\cos 2t$

b $\dfrac{1}{2}$ **c** $y + 2x = \dfrac{5\sqrt{2}}{2}$

5 a $\dfrac{dy}{dt} = 3t^2 - 4$, $\dfrac{dx}{dt} = 2$, $\dfrac{dy}{dx} = \dfrac{3t^2 - 4}{2}$

At $t = -1$, $\dfrac{dy}{dx} = -\dfrac{1}{2}$, $x = 1$, $y = 3$.

Equation of l is $2y + x = 7$.

b 2

6 $\dfrac{dV}{dt} = -kV$ **7** $\dfrac{dM}{dt} = -kM$ **8** $\dfrac{dP}{dt} = kP - Q$

9 $\dfrac{dr}{dt} = \dfrac{k}{r}$ **10** $\dfrac{d\theta}{dt} = -k(\theta - \theta_0)$

11 a $\dfrac{\pi}{6}$ **b** $-\dfrac{3}{16}\cosec t$

c $\dfrac{dy}{dx} = -\dfrac{3}{8} \Rightarrow$ gradient of normal $= \dfrac{8}{3}$

$y - \dfrac{3}{2} = \dfrac{8}{3}(x - 2) \Rightarrow 6y - 16x + 23 = 0$

d $-\dfrac{123}{64}$

12 a $-\dfrac{1}{2}\sec t$ **b** $4y + 4x = 5a$

c Tangent crosses the x-axis at $x = \dfrac{5}{4}a$, and crosses the y-axis at $y = \dfrac{5}{4}a$.

So area $AOB = \dfrac{1}{2}\left(\dfrac{5}{4}a\right)^2 = \dfrac{25}{32}a^2$, $k = \dfrac{25}{32}$

13 $y + x = 16$ **14** $\dfrac{1}{7}$

15 $\dfrac{y - 2e^{2x}}{2e^{2y} - x}$ **16** $(1, 1)$ and $(-\sqrt[3]{3}, \sqrt[3]{-3})$.

17 a $\dfrac{2x - 2 - y}{1 + x - 2y}$ **b** $\dfrac{4}{3}, -\dfrac{1}{3}$

c $\left(\dfrac{5 + 2\sqrt{13}}{3}, \dfrac{4 + \sqrt{13}}{3}\right)$ and $\left(\dfrac{5 - 2\sqrt{13}}{3}, \dfrac{4 - \sqrt{13}}{3}\right)$

18 $14x + 48y + 48x\dfrac{dy}{dx} - 14y\dfrac{dy}{dx} = 0 \Rightarrow \dfrac{dy}{dx} = \dfrac{-7x - 24y}{24x - 7y}$

So $\dfrac{-7x - 24y}{24x - 7y} = \dfrac{2}{11} \Rightarrow -77x - 264y$

$\qquad\qquad\qquad = 48x - 14y \Rightarrow x + 2y = 0$

19 $\ln y = x\ln x \Rightarrow \dfrac{1}{y} \times \dfrac{dy}{dx} = x\dfrac{d}{dx}(\ln x) + \dfrac{d}{dx}(x)\ln x = 1 + \ln x$

So $\dfrac{dy}{dx} = y(1 + \ln x) = x^x(1 + \ln x)$

20 a $\ln a^x = \ln e^{kx} \Rightarrow x\ln a = kx\ln e = kx \Rightarrow k = \ln a$

   b $y = e^{(\ln 2)x} \Rightarrow \dfrac{dy}{dx} = \ln 2\, e^{(\ln 2)x} = 2^x \ln 2$

   c $\dfrac{dy}{dx} = 2^2 \ln 2 = 4\ln 2 = \ln 2^4 = \ln 16$

21 a $\dfrac{\ln P - \ln P_0}{\ln 1.09}$ **b** 8.04 years   **c** $0.172P_0$

22 a $\left(\dfrac{\pi}{2}, 0\right)$

   b $\dfrac{d^2y}{dx^2} = -\mathrm{cosec}^2x.\ \mathrm{cosec}^2x > 0$ for all x,

      hence $-\mathrm{cosec}^2x < 0$, so $\dfrac{d^2y}{dx^2} < 0$ for all x.

      Thus C is concave for all values of x.

23 a $40e^{-0.183} = 33.31\ldots$ **b** $-9.76e^{-0.244t}$

   c The mass is decreasing

24 a $f'(x) = -\dfrac{2\sin 2x + \cos 2x}{e^x}$

      $f'(x) = 0 \Leftrightarrow 2\sin 2x + \cos 2x = 0 \Leftrightarrow \tan 2x = -0.5$

      $A(1.34, -0.234), B(2.91, 0.0487)$

   b Maximum (6.91, 2.19); minimum (5.34, 1.06) to 3 s.f.

   c $0 < x \leqslant 0.322, 1.89 \leqslant x < \pi$

Challenge

a $-\dfrac{4\cos 2t}{5\sin\left(t + \dfrac{\pi}{12}\right)}$

b $\left(\dfrac{5}{2}, 2\right), \left(-\dfrac{5\sqrt{3}}{2}, -2\right), \left(-\dfrac{5}{2}, 2\right), \left(\dfrac{5\sqrt{3}}{2}, -2\right)$

c Cuts the x–axis at:
  (4.83, 0) gradient -3.09; $(-1.29, 0)$ gradient 0.828
  $(-4.83, 0)$ gradient 3.09; (1.29, 0) gradient -0.828
  Cuts the y–axis twice at (0, 1) gradients 0.693 and -0.693

d $(-5, -1)$ and $(5, -1)$

e

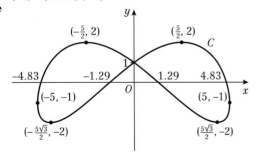

CHAPTER 6

Prior knowledge check

1 a $12(2x - 7)^5$ **b** $5\cos 5x$ **c** $\dfrac{1}{3}e^{\frac{x}{3}}$

2 a $y = \dfrac{16}{3}x^{\frac{3}{2}} - 12x^{\frac{1}{2}}$ **b** $\dfrac{268}{3}$

3 $\dfrac{7}{4x - 1} - \dfrac{1}{x + 3}$ **4** 6 units^2

Exercise 6A

1 9

2 16

3 $\dfrac{\pi}{3} + \dfrac{\sqrt{3}}{4}$

4 $b = 5$

5 $\ln 256$

6 65050

Exercise 6B

1 a 640π **b** $\dfrac{8}{3\pi}$ **c** 48π **d** $\dfrac{55}{24}\pi$

2 250π

3 $\dfrac{141\pi}{10}$

4 8π

5 a (3, 0)   **b** $\dfrac{2187}{20}\pi$

6 5.97

7 $\dfrac{279\pi}{20}$

8 $a = \dfrac{1}{2}$

9 $y = r,\ \pi\int_0^h r^2 dx = \pi[r^2 x]_0^h = \pi r^2 h$

10 $\dfrac{3}{4}\pi\,(9\sqrt{3} - 1)$ or 27.5

11 $18\pi\ln\left(\dfrac{3}{2}\right)$

Challenge

$\dfrac{35}{2}\pi$

Exercise 6C

1 a $\dfrac{2}{5}(1 + x)^{\frac{5}{2}} - \dfrac{2}{3}(1 + x)^{\frac{3}{2}} + c$ **b** $-\ln|1 - \sin x| + c$

   c $\dfrac{\cos^3 x}{3} - \cos x + c$ **d** $\ln\left|\dfrac{\sqrt{x} - 2}{\sqrt{x} + 2}\right| + c$

   e $\dfrac{2}{5}(1 + \tan x)^{\frac{5}{2}} - \dfrac{2}{3}(1 + \tan x)^{\frac{3}{2}} + c$

   f $\tan x + \dfrac{1}{3}\tan^3 x + c$

2 a $\dfrac{506}{15}$ **b** $\dfrac{392}{5}$ **c** $\dfrac{14}{9}$ **d** $\dfrac{16}{3} - 2\sqrt{3}$ **e** $\dfrac{1}{2}\ln\dfrac{9}{5}$

3 a $\dfrac{(3 + 2x)^7}{28} - \dfrac{(3 + 2x)^6}{8} + c$ **b** $\dfrac{2}{3}(1 + x)^{\frac{3}{2}} - 2\sqrt{1 + x} + c$

   c $\sqrt{x^2 + 4} + \ln\left|\dfrac{\sqrt{x^2 + 4} - 2}{\sqrt{x^2 + 4} + 2}\right| + c$

4 a $\dfrac{886}{15}$ **b** $2 + 2\ln\dfrac{2}{3}$ **c** $2 - 2\ln 2$

5 $\dfrac{592}{3}$

6 $\displaystyle\int_{\ln 3}^{\ln 4} \dfrac{e^{4x}}{e^x - 2}\,dx = \int_1^{\sqrt{2}} \dfrac{2(u^2 + 2)^3}{u}\,du$

   $\displaystyle = \int_1^{\sqrt{2}}\left(2u^5 + 12u^3 + 24u + \dfrac{16}{u}\right)du$

   $\displaystyle = \left[\dfrac{1}{3}u^6 + 3u^4 + 12u^2 + 16\ln u\right]_1^{\sqrt{2}}$

   $\displaystyle = \left(\dfrac{116}{3} + 16\ln\sqrt{2}\right) - \left(\dfrac{46}{3} + 16\ln 1\right)$

   $= \dfrac{70}{3} + 16\ln\sqrt{2} = \dfrac{70}{3} + 8\ln 2 \Rightarrow$

   $a = 70, b = 3, c = 8, d = 2$

7 $\int_0^{\frac{\pi}{3}} \sin^3 x \cos^2 x \, dx = \int_1^{\frac{1}{2}}(u^2-1)u^2 \, du = \int_1^{\frac{1}{2}}(u^4-u^2) \, du$

$\qquad\qquad = \left[\frac{1}{5}u^5 - \frac{1}{3}u^3\right]_1^{\frac{1}{2}} = \frac{47}{480}$

8 $\dfrac{2\pi + 3\sqrt{3}}{96}$

Challenge

$x = 3\sin u, \dfrac{dx}{du} = 3\cos u \Rightarrow dx = 3\cos u \, du$

$(3\sin u)^2 + (3\cos u)^2 = 9$

$\Rightarrow x^2 + (3\cos u)^2 = 9 \Rightarrow \cos u = \dfrac{\sqrt{9-x^2}}{3}$

$\int \dfrac{1}{x^2\sqrt{9-x^2}} \, dx = \int \dfrac{1}{9\sin^2 u \cos u}(3\cos u) \, du$

$\qquad\qquad = \dfrac{1}{9}\int \operatorname{cosec}^2 u \, du = -\dfrac{1}{9}\cot u + c = -\dfrac{\cos u}{9\sin u} + c$

$\qquad\qquad = -\dfrac{\frac{\sqrt{9-x^2}}{3}}{\frac{3x}{3}} + c = -\dfrac{\sqrt{9-x^2}}{9x} + c$

Exercise 6D

1 **a** $-x\cos x + \sin x + c$ **b** $xe^x - e^x + c$
  c $x\tan x - \ln|\sec x| + c$ **d** $x\sec x - \ln|\sec x + \tan x| + c$
  e $-x\cot x + \ln|\sin x| + c$

2 **a** $3x\ln x - 3x + c$ **b** $\dfrac{x^2}{2}\ln x - \dfrac{x^2}{4} + c$
  c $-\dfrac{\ln x}{2x^2} - \dfrac{1}{4x^2} + c$ **d** $x(\ln x)^2 - 2x\ln x + 2x + c$
  e $\dfrac{x^3}{3}\ln x - \dfrac{x^3}{9} + x\ln x - x + c$

3 **a** $-e^{-x}x^2 - 2xe^{-x} - 2e^{-x} + c$
  b $x^2\sin x + 2x\cos x - 2\sin x + c$
  c $x^2(3+2x)^6 - \dfrac{x(3+2x)^7}{7} + \dfrac{(3+2x)^8}{112} + c$
  d $-x^2\cos 2x + x\sin 2x + \dfrac{1}{2}\cos 2x + c$
  e $x^2\sec^2 x - 2x\tan x + 2\ln|\sec x| + c$

4 **a** $2\ln 2 - \dfrac{3}{4}$ **b** 1 **c** $\dfrac{\pi}{2} - 1$
  d $\dfrac{1}{2}(1 - \ln 2)$ **e** 9.8 **f** $2\sqrt{2}\pi + 8\sqrt{2} - 16$
  g $\dfrac{1}{2}(1 - \ln 2)$

5 **a** $\dfrac{1}{16}(4x\sin 4x + \cos 4x) + c$
  b $\dfrac{1}{32}((1 - 8x^2)\cos 4x + 4x\sin 4x) + c$

6 **a** $-\dfrac{2}{3}(8-x)^{\frac{3}{2}} + c$
  b $u = x - 2 \Rightarrow \dfrac{du}{dx} = 1; \dfrac{dv}{dx} = \sqrt{8-x} \Rightarrow v = -\dfrac{2}{3}(8-x)^{\frac{3}{2}}$

  $I = (x-2)\left(-\dfrac{2}{3}(8-x)^{\frac{3}{2}}\right) - \int -\dfrac{2}{3}(8-x)^{\frac{3}{2}} \, dx$

  $\quad = -\dfrac{2}{3}(x-2)(8-x)^{\frac{3}{2}} + \dfrac{2}{3}\int (8-x)^{\frac{3}{2}} \, dx$

  $\quad = -\dfrac{2}{3}(x-2)(8-x)^{\frac{3}{2}} - \dfrac{4}{15}(8-x)^{\frac{5}{2}} + c$

  $\quad = -\dfrac{2}{3}(x-2)(8-x)^{\frac{3}{2}} - \dfrac{4}{15}(8-x)(8-x)^{\frac{3}{2}} + c$

  $\quad = (8-x)^{\frac{3}{2}}\left(-\dfrac{2}{3}(x-2) - \dfrac{4}{15}(8-x)\right) + c$

  $\quad = (8-x)^{\frac{3}{2}}\left(-\dfrac{2x}{5} + \dfrac{4}{5}\right) + c = -\dfrac{2}{5}(8-x)^{\frac{3}{2}}(x+2) + c$

c 15.6

7 **a** $\dfrac{1}{3}\tan 3x + c$ **b** $\dfrac{1}{3}x\tan 3x - \dfrac{1}{9}\ln|\sec 3x| + c$

  c $\int_{\frac{\pi}{18}}^{\frac{\pi}{9}} x\sec^2 3x = \left[\dfrac{1}{3}x\tan 3x - \dfrac{1}{9}\ln|\sec 3x|\right]_{\frac{\pi}{18}}^{\frac{\pi}{9}}$

  $\qquad = \left(\dfrac{\sqrt{3}\pi}{27} - \dfrac{1}{9}\ln 2\right) - \left(\dfrac{\sqrt{3}\pi}{162} - \dfrac{1}{9}\ln\dfrac{2}{\sqrt{3}}\right)$

  $\qquad = \dfrac{5\sqrt{3}\pi}{162} - \dfrac{1}{9}\ln 2 + \dfrac{1}{9}\ln 2 - \dfrac{1}{9}\ln\sqrt{3}$

  $\qquad = \dfrac{5\sqrt{3}\pi}{162} - \dfrac{1}{18}\ln 3 \Rightarrow p = \dfrac{5\sqrt{3}}{162}$ and $q = \dfrac{1}{18}$

Exercise 6E

1 **a** $\ln|(x+1)^2(x+2)| + c$ **b** $\ln|(x-2)\sqrt{2x+1}| + c$
  c $\ln\left|\dfrac{(x+3)^3}{x-1}\right| + c$ **d** $\ln\left|\dfrac{2+x}{1-x}\right| + c$

2 **a** $x + \ln|(x+1)^2\sqrt{2x-1}| + c$ **b** $\dfrac{x^2}{2} + x + \ln\left|\dfrac{x^2}{(x+1)^3}\right| + c$
  c $x + \ln\left|\dfrac{x-2}{x+2}\right| + c$ **d** $-x + \ln\left|\dfrac{(3+x)^2}{1-x}\right| + c$

3 **a** $A = 2, B = 2$ **b** $\ln\left|\dfrac{2x+1}{1-2x}\right| + c$ **c** $\ln\dfrac{5}{9}$, so $k = \dfrac{5}{9}$

4 **a** $f(x) = \dfrac{2}{3+2x} + \dfrac{1}{2-x} + \dfrac{1}{(2-x)^2}$ **b** $\dfrac{1}{2} + \ln\dfrac{10}{3}$

5 **a** $A = 1, B = 2, C = -2$ **b** $a = \dfrac{2}{3}, b = -\dfrac{4}{3}, c = 3$

6 **a** $f(x) = \dfrac{3}{x^2} - \dfrac{1}{x+2}$ **b** $a = \dfrac{3}{4}, b = \dfrac{2}{3}$

7 **a** $f(x) = 2 - \dfrac{3}{4x+1} + \dfrac{3}{4x-1}$, so $A = 2, B = -3$ and $C = 3$
  b $k = \dfrac{3}{4}, m = \dfrac{35}{27}$

Exercise 6F

1 **a** $y = Ae^{x-x^2} - 1$ **b** $y = k\sec x$
  c $y = \dfrac{-1}{\tan x - x + c}$ **d** $y = \ln|2e^x + c|$

2 **a** $\dfrac{1}{24} - \dfrac{\cos^3 x}{3}$ **b** $\sin 2y + 2y = 4\tan x - 4$
  c $\tan y = \dfrac{1}{2}\sin 2x + x + 1$ **d** $y = \arccos(e^{-\tan x})$

3 **a** $y = Axe^{-\frac{1}{x}}$ **b** $y = -e^3 xe^{-\frac{1}{x}} = -xe^{(\frac{3x-1}{x})}$

4 $y = \sqrt{\dfrac{x}{x+1}}$ **5** $\ln|2+e^y| = -xe^{-x} - e^{-x} + c$

6 $y = \dfrac{3}{1-x}$ **7** $y = \dfrac{3(1+x^2)+1}{3(1+x^2)-1}$

8 $y = \ln\left|\dfrac{x^2-12}{2}\right|$ **9** $\tan y = x + \dfrac{1}{2}\sin 2x + \dfrac{2-\pi}{4}$

10 $\ln|y| = -x\cos x + \sin x - 1$

11 **a** $3x + 4\ln|x| + c$ **b** $y = \left(\dfrac{3}{2}x + 2\ln|x| + \dfrac{5}{2}\right)^2$

12 **a** $\dfrac{5}{3x-8} + \dfrac{1}{x-2}$
  b $\ln|y| = \dfrac{5}{3}\ln|3x-8| + \ln|x-2| + c$
  c $y = 8(x-2)(3x-8)^{\frac{5}{3}}$

13 **a** $y = x^2 - 4x + c$
  b Graphs of the form $y = x^2 - 4x + c$, where c is any real number

14 a $y = \dfrac{1}{x+2} + c$

b Graphs of the form $y = \dfrac{1}{x+2} + c$, where c is any real number

c $y = \dfrac{1}{x+2} + 3$

15 a $\dfrac{dy}{dx} = -\dfrac{x}{y} \Rightarrow \int y \, dy = \int -x \, dx$

$\Rightarrow \dfrac{1}{2}y^2 = -\dfrac{1}{2}x^2 + b \Rightarrow y^2 + x^2 = c$

b Circles with centre $(0, 0)$ and radius $\sqrt{c}$, where c is any positive real number.

c $y^2 + x^2 = 49$

Exercise 6G

1 a $y = 200e^{kt}$ **b** 1 year

c The population could not increase in size in this way forever due to limitations such as available food or space

2 a $M = \dfrac{e^t}{1 + e^t}$ **b** $\dfrac{2}{3}$ **c** M approaches 1

3 a $\dfrac{dx}{dt} = \dfrac{k}{x^2} \Rightarrow \dfrac{1}{3}x^3 = kt + c$

$t = 0, x = 1 \Rightarrow c = \dfrac{1}{3} \Rightarrow t = 20, x = 2 \Rightarrow k = \dfrac{7}{60}$

$\dfrac{1}{3}x^3 = \dfrac{7}{60}t + \dfrac{1}{3} \Rightarrow x = \sqrt[3]{\left(\dfrac{7}{20}t + 1\right)}$

b $x = 3, t = 74.3$ days. So it takes 54.3 days to increase from 2 cm to 3 cm.

4 a The difference in temperature is $T - 25$. The tea is cooling, so there should be a negative sign. k has to be positive or the tea would be warming.

b 46.2°C

5 a $\int A^{-\frac{3}{2}} dA = \dfrac{1}{10}\int t^{-2} dt \Rightarrow \dfrac{-2}{\sqrt{A}} = \dfrac{-1}{10t} + C \Rightarrow C = -\dfrac{19}{10}$

$\Rightarrow \dfrac{-2}{\sqrt{A}} = \dfrac{-1}{10t} - \dfrac{19}{10} \Rightarrow \sqrt{A} = \left(\dfrac{-20t}{-1 - 19t}\right)$

$\Rightarrow A = \left(\dfrac{20t}{1 + 19t}\right)^2$

b As $t \to \infty, A \to \left(\dfrac{20}{19}\right)^2 = \dfrac{400}{361}$ from below

6 a $V = 6000h \Rightarrow \dfrac{dV}{dh} = 6000, \dfrac{dV}{dt} = 12000 - 500h,$

$\dfrac{dh}{dt} = \dfrac{dV}{dt} \div \dfrac{dV}{dh} = \dfrac{1}{6000}(12000 - 500h)$

$60\dfrac{dh}{dt} = 120 - 5h$

b $t = 12\ln\left(\dfrac{9}{7}\right)$

7 a $\dfrac{\left(\dfrac{1}{10000}\right)}{P} + \dfrac{\left(\dfrac{1}{10000}\right)}{10000 - P}$

b $P = \dfrac{10000}{1 + 3e^{-50t}}$ so $a = 10000, b = 1$ and $c = 3$.

c 10 000 deer

8 a $\dfrac{dV}{dt} = 40 - \dfrac{1}{4}V \Rightarrow -4\dfrac{dV}{dt} = V - 160$

b $V = 160 + 4840e^{-\frac{1}{4}t}, a = 160$ and $b = 4840$

c $V \to 160$

9 a $\dfrac{dR}{dt} = -kR \Rightarrow \int\dfrac{1}{R}dR = -k\int dt$

$\Rightarrow \ln R = -kt + c \Rightarrow R = e^{-kt+c}$

$\Rightarrow R = Ae^{-kt} \Rightarrow R_0 = Ae^0 \Rightarrow A = R_0 \Rightarrow R = R_0e^{-kt}$

b $k = \dfrac{1}{5730}\ln 2$

c $0.1R_0 = R_0e^{\frac{1}{5730}\ln\left(\frac{1}{2}\right)\times t}$

$\ln(0.1) = \dfrac{1}{5730}\ln\left(\dfrac{1}{2}\right) \times t \Rightarrow t \approx 19035$

Chapter review 6

1 8.8

2 $40 - 8\ln 3$

3 $\dfrac{4374}{35}\pi$

4 a $\dfrac{1}{4}$ **b** $\dfrac{81}{8}\pi$

5 a $x^2 + 4x + 4 = y \Rightarrow y = (x + 2)^2$

$\Rightarrow x = \sqrt{y} - 2 \Rightarrow x^2 = 4 - 4\sqrt{y} + y$

b $\dfrac{11}{6}\pi$

6 $\dfrac{56\pi}{5}$

7 a $\dfrac{2}{4}(2 + 1)^2 = 4.5$

$3 \times 2 + 4 \times 4.5 = 24$

b $\dfrac{1549}{210}\pi + \dfrac{81}{2}\pi = \dfrac{5027}{105}\pi$

8 $\dfrac{28}{3}\pi$

9 0.356

10 $\pi\ln 16$

11 a $\dfrac{1}{40}(4x - 1)^{\frac{5}{2}} + \dfrac{1}{24}(4x - 1)^{\frac{3}{2}} + c$

b $\dfrac{x^2}{2}\ln x - \dfrac{1}{4}x^2 + c$

c $-\dfrac{1}{4}\ln|\cos 2x| + c$

12 a $\dfrac{1}{4}\pi - \dfrac{1}{2}\ln 2$ **b** $\dfrac{1}{4}\ln\left(\dfrac{35}{19}\right)$ **c** $\ln\left(\dfrac{4}{3}\right)$

13 a $\int\dfrac{1}{x^2}\ln x \, dx = (\ln x)\left(-\dfrac{1}{x}\right) - \int\left(-\dfrac{1}{x}\right)\left(\dfrac{1}{x}\right)dx$

$= -\dfrac{\ln x}{x} + \int\dfrac{1}{x^2}dx = -\dfrac{\ln x}{x} - \dfrac{1}{x} + c$

$-\dfrac{2}{e}\int_1^e\dfrac{1}{x^2}\ln x \, dx = \left[-\dfrac{\ln x}{x} - \dfrac{1}{x}\right]_1^e = \left(-\dfrac{1}{e} - \dfrac{1}{e}\right) - (0 - 1) = 1$

b $\dfrac{1}{(x + 1)(2x - 1)} = \dfrac{A}{x + 1} + \dfrac{B}{2x - 1} \Rightarrow A = -\dfrac{1}{3}, B = \dfrac{2}{3}$

$\int_1^p\dfrac{1}{(x + 1)(2x - 1)}dx = \int_1^p\left(-\dfrac{1}{3(x + 1)} + \dfrac{2}{3(2x - 1)}\right)dx$

$= \left[-\dfrac{1}{3}\ln(x + 1) + \dfrac{1}{3}\ln(2x - 1)\right]_1^p = \left[\dfrac{1}{3}\ln\left(\dfrac{2x - 1}{x + 1}\right)\right]_1^p$

$= \left(\dfrac{1}{3}\ln\left(\dfrac{2p - 1}{p + 1}\right)\right) - \left(\dfrac{1}{3}\ln\left(\dfrac{1}{2}\right)\right)$

$= \dfrac{1}{3}\ln\left(\dfrac{2(2p - 1)}{p + 1}\right) = \dfrac{1}{3}\ln\left(\dfrac{4p - 2}{p + 1}\right)$

14 a $\dfrac{2}{3}(x - 2)\sqrt{x + 1} + c$ **b** $\dfrac{8}{3}$

15 a $-\dfrac{1}{8}x\cos 8x + \dfrac{1}{64}\sin 8x + c$

b $\dfrac{1}{8}x^2\sin 8x + \dfrac{1}{32}x\cos 8x - \dfrac{1}{256}\sin 8x + c$

16 a $A = \dfrac{1}{2}, B = 2, C = -1$

b $\dfrac{1}{2}\ln|x| + 2\ln|x - 1| + \dfrac{1}{x - 1} + c$

Online Worked solutions are available in SolutionBank.

c $\int_4^9 f(x)\,dx = \left[\frac{1}{2}\ln|x| + 2\ln|x-1| + \frac{1}{x-1}\right]_4^9$

$= \left(\frac{1}{2}\ln 9 + 2\ln 8 + \frac{1}{8}\right) - \left(\frac{1}{2}\ln 4 + 2\ln 3 + \frac{1}{3}\right)$

$= \left(\ln 3 + \ln 64 + \frac{1}{8}\right) - \left(\ln 2 + \ln 9 + \frac{1}{3}\right)$

$= \ln\left(\frac{3 \times 64}{2 \times 9}\right) - \frac{5}{24} = \ln\left(\frac{32}{3}\right) - \frac{5}{24}$

17 a $\frac{1}{3}x^3 \ln 2x - \frac{1}{9}x^3 + c$

b $\left[\frac{1}{3}x^3 \ln 2x - \frac{1}{9}x^3\right]_{\frac{1}{2}}^{3} = (9\ln 6 - 3) - \left(0 - \frac{1}{72}\right)$

$= 9\ln 6 - \frac{215}{72}$

18 a $-x\,e^{-x} - e^{-x} + c$　　　　**b** $\cos 2y = 2e^{-x}(x - e^x + 1)$

19 a $-\frac{1}{2}x\cos 2x + \frac{1}{4}\sin 2x + c$

b $\tan y = -\frac{1}{2}x\cos 2x + \frac{1}{4}\sin 2x - \frac{1}{4}$

20 a $-\frac{1}{y} = \frac{1}{2}x^2 + c$

b $x = 1: -\frac{1}{1} = \frac{1}{2} + c \Rightarrow c = -\frac{3}{2}$

$-\frac{1}{y} = \frac{1}{2}x^2 - \frac{3}{2} \Rightarrow \frac{1}{y} = \frac{1}{2}(3 - x^2) \Rightarrow y = \frac{2}{3 - x^2}$

c 1　　　　**d** $y = x;\ (-2, -2)$

21 a $\ln|1 + 2x| + \frac{1}{1 + 2x} + c$

b $2y - \sin 2y = \ln|1 + 2x| + \frac{1}{1 + 2x} + \frac{\pi}{2} - 2$

22 a $A_1 = \frac{1}{4} - \frac{1}{2e}, A_2 = \frac{1}{4}$

b $A_1 : A_2 \Rightarrow \frac{1}{4} - \frac{1}{2e} : \frac{1}{4} \Rightarrow 1 - \frac{2}{e} : 1 \Rightarrow (e - 2) : e$

23 a $-e^{-x}(x^2 + 2x + 2) + c$

b $y = -\frac{1}{3}\ln|3e^{-x}(x^2 + 2x + 2) - 5|$

24 a $A = 1, B = \frac{1}{2}, C = -\frac{1}{2}$

b $x + \frac{1}{2}\ln|x - 1| - \frac{1}{2}\ln|x + 1| = 2t - \frac{1}{2}\ln 3$

25 a $\frac{dV}{dt} = -kV \Rightarrow \int\frac{1}{V}\,dV = \int -k\,dt \Rightarrow \ln V = -kt + c$

$\Rightarrow V = Ae^{-kt}$

b

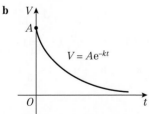

c $\frac{1}{2}A = Ae^{-kT} \Rightarrow \frac{1}{2} = e^{-kT} \Rightarrow 2 = e^{kT} \Rightarrow \ln 2 = kT$

26 a $\int(k - y)\,dy = \int x\,dx \Rightarrow ky - \frac{1}{2}y^2 = \frac{1}{2}x^2 + c$

$x^2 + (y - k)^2 = C$

b Concentric circles with centre (0, 2).

27 a $\frac{(1 + 2x^2)^6}{24} + c$　　　**b** $\tan 2y = \frac{1}{12}(1 + 2x^2)^6 + \frac{11}{12}$

28 $\arctan x + c$　　　**29** $y^2 = \frac{8x}{x + 2}$

30 a $A = \pi r^2 \Rightarrow \frac{dA}{dr} = 2\pi r$

$\frac{dr}{dt} = \frac{dr}{dA} \times \frac{dA}{dt} = \frac{1}{2\pi r} \times k\sin\left(\frac{t}{3\pi}\right) = \frac{k}{2\pi r}\sin\left(\frac{t}{3\pi}\right)$

b $r^2 = -6\cos\left(\frac{t}{3\pi}\right) + 7$　　　**c** 6.19 days

Challenge
a 15　　　**b** −3

CHAPTER 7
Prior knowledge check

1 a $\begin{pmatrix}4\\2\end{pmatrix}$　　**b** $\begin{pmatrix}5\\-2\end{pmatrix}$　　**c** $\begin{pmatrix}-1\\-3\end{pmatrix}$

2 a $\frac{7}{9}$　　**b** $\frac{2}{9}$　　**c** $\frac{7}{2}$

3 a 123.2°　　**b** 13.6　　**c** 5.3　　**d** 21.4°

Exercise 7A

1 a 　　　**b**

c 　　　**d**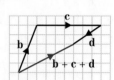

e 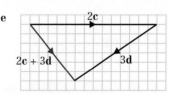　　　**f**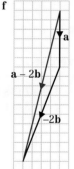

g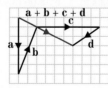

2 a 2**b**　　　**b** **d**　　　**c** **b**
d 2**b**　　　**e** **d** + **b**　　**f** **d** + **b**
g −2**d**　　　**h** −**b**　　　**i** 2**d** + **b**
j −**b** + 2**d**　**k** −**b** + **d**　**l** −**d** − **b**

3 a 2**m**　　　**b** 2**p**　　　**c** **m**
d **m**　　　**e** **p** + **m**　　**f** **p** + **m**
g **p** + 2**m**　**h** **p** − **m**　　**i** −**m** − **p**
j −2**m** + **p**　**k** −2**p** + **m**　**l** −**m** − 2**p**

4 a **d** − **a**　　**b** **a** + **b** + **c**
c **a** + **b** − **d**　**d** **a** + **b** + **c** − **d**

5 a 2**a** + 2**b**　**b** **a** + **b**　　**c** **b** − **a**

6 a **b**　　　　**b** **b** − 3**a**　　**c** **a** − **b**
d 2**a** − **b**

7 a $\overrightarrow{OB} = \mathbf{a} + \mathbf{b}$　**b** $\overrightarrow{OP} = \frac{5}{8}(\mathbf{a} + \mathbf{b})$　**c** $\overrightarrow{AP} = \frac{5}{8}\mathbf{b} - \frac{3}{8}\mathbf{a}$

8 a Yes ($\lambda = 2$)　**b** Yes ($\lambda = 4$)　**c** No
d Yes ($\lambda = -1$)　**e** Yes ($\lambda = -3$)　**f** No

9 a i $\mathbf{b} - \mathbf{a}$ **ii** $\frac{1}{2}\mathbf{a}$ **iii** $\frac{1}{2}\mathbf{b}$ **iv** $\frac{1}{2}\mathbf{b} - \frac{1}{2}\mathbf{a}$

  b $\overrightarrow{BC} = \mathbf{b} - \mathbf{a}$, $\overrightarrow{PQ} = \frac{1}{2}(\mathbf{b} - \mathbf{a})$ so PQ is parallel to BC.

10 a i $2\mathbf{b}$ **ii** $\mathbf{a} - \mathbf{b}$

  b $\overrightarrow{AB} = 2\mathbf{b}$, $\overrightarrow{OC} = 3\mathbf{b}$ so AB is parallel to OC.

11 1.2

Exercise 7B

1 $\mathbf{v_1}$: $8\mathbf{i}$, $\begin{pmatrix} 8 \\ 0 \end{pmatrix}$ $\mathbf{v_2}$: $9\mathbf{i} + 3\mathbf{j}$, $\begin{pmatrix} 9 \\ 3 \end{pmatrix}$ $\mathbf{v_3}$: $-4\mathbf{i} + 2\mathbf{j}$, $\begin{pmatrix} -4 \\ 2 \end{pmatrix}$

  $\mathbf{v_4}$: $3\mathbf{i} + 5\mathbf{j}$, $\begin{pmatrix} 3 \\ 5 \end{pmatrix}$ $\mathbf{v_5}$: $-3\mathbf{i} - 2\mathbf{j}$, $\begin{pmatrix} -3 \\ -2 \end{pmatrix}$ $\mathbf{v_6}$: $-5\mathbf{j}$, $\begin{pmatrix} 0 \\ -5 \end{pmatrix}$

2 a $8\mathbf{i} + 12\mathbf{j}$ **b** $\mathbf{i} + 1.5\mathbf{j}$ **c** $-4\mathbf{i} + \mathbf{j}$
  d $10\mathbf{i} + \mathbf{j}$ **e** $-2\mathbf{i} + 11\mathbf{j}$ **f** $-2\mathbf{i} - 10\mathbf{j}$
  g $14\mathbf{i} - 7\mathbf{j}$ **h** $-8\mathbf{i} + 9\mathbf{j}$

3 a $\begin{pmatrix} 45 \\ 35 \end{pmatrix}$ **b** $\begin{pmatrix} 4 \\ 0.5 \end{pmatrix}$ **c** $\begin{pmatrix} 12 \\ 3 \end{pmatrix}$

  d $\begin{pmatrix} -1 \\ 16 \end{pmatrix}$ **e** $\begin{pmatrix} -21 \\ -29 \end{pmatrix}$ **f** $\begin{pmatrix} 10 \\ 2 \end{pmatrix}$

4 a $\lambda = 5$ **b** $\mu = -\frac{3}{2}$

5 a $\lambda = \frac{1}{3}$ **b** $\mu = -1$

  c $s = -1$ **d** $t = -\frac{1}{17}$

6 $\mathbf{i} - \mathbf{j}$

7 a $\overrightarrow{AC} = 5\mathbf{i} - 4\mathbf{j} = \begin{pmatrix} 5 \\ -4 \end{pmatrix}$ **b** $\overrightarrow{AP} = 3\mathbf{i} - \frac{12}{5}\mathbf{j} = \begin{pmatrix} 3 \\ -\frac{12}{5} \end{pmatrix}$

  c $\overrightarrow{OP} = 5\mathbf{i} + \frac{8}{5}\mathbf{j} = \begin{pmatrix} 5 \\ \frac{8}{5} \end{pmatrix}$

8 $j = 4$, $k = 11$

9 $p = 3$, $q = 2$

10 a $p = 5$ **b** $8\mathbf{i} - 12\mathbf{j}$

Exercise 7C

1 a 5 **b** 10 **c** 13
  d 4.47 (3 s.f.) **e** 5.83 (3 s.f.) **f** 8.06 (3 s.f.)
  g 5.83 (3 s.f.) **h** 4.12 (3 s.f.)

2 a $\sqrt{26}$ **b** $5\sqrt{2}$ **c** $\sqrt{101}$

3 a $\frac{1}{5}\begin{pmatrix} 4 \\ 3 \end{pmatrix}$ **b** $\frac{1}{13}\begin{pmatrix} 5 \\ -12 \end{pmatrix}$

  c $\frac{1}{25}\begin{pmatrix} -7 \\ 24 \end{pmatrix}$ **d** $\frac{1}{\sqrt{10}}\begin{pmatrix} 1 \\ -3 \end{pmatrix}$

4 a 53.1° above **b** 53.1° below
  c 67.4° above **d** 63.4° above

5 a 149° to the right **b** 29.7° to the right
  c 31.0° to the left **d** 104° to the left

6 a $\frac{15\sqrt{2}}{2}\mathbf{i} + \frac{15\sqrt{2}}{2}\mathbf{j}$, $\begin{pmatrix} \frac{15\sqrt{2}}{2} \\ \frac{15\sqrt{2}}{2} \end{pmatrix}$ **b** $7.52\mathbf{i} + 2.74\mathbf{j}$, $\begin{pmatrix} 7.52 \\ 2.74 \end{pmatrix}$

  c $18.1\mathbf{i} - 8.45\mathbf{j}$, $\begin{pmatrix} 18.1 \\ -8.45 \end{pmatrix}$ **d** $\frac{5\sqrt{3}}{2}\mathbf{i} - 2.5\mathbf{j}$, $\begin{pmatrix} \frac{5\sqrt{3}}{2} \\ -2.5 \end{pmatrix}$

7 a $|3\mathbf{i} + 4\mathbf{j}| = 5$, 53.1° above

  b $|2\mathbf{i} - \mathbf{j}| = \sqrt{5}$, 26.6° below

  c $|-5\mathbf{i} + 2\mathbf{j}| = \sqrt{29}$, 158.2° above

8 $k = \pm 6$

9 $p = \pm 8$, $q = 6$

10 a 36.9° **b** 33.7° **c** 70.6°

11 a 67.2° **b** 19.0

Exercise 7D

1 a i $\begin{pmatrix} -3 \\ 5 \\ -9 \end{pmatrix}$ **ii** $\begin{pmatrix} 11 \\ -11 \\ 19 \end{pmatrix}$

  b $\mathbf{a} - \mathbf{b}$ is parallel as $-2(\mathbf{a} - \mathbf{b}) = 6\mathbf{i} - 10\mathbf{j} + 18\mathbf{k}$
  $-\mathbf{a} + 3\mathbf{b}$ is not parallel as it is not a multiple of
  $6\mathbf{i} - 10\mathbf{j} + 18\mathbf{k}$.

2 $3\mathbf{a} + 2\mathbf{b} = 3\begin{pmatrix} 3 \\ 2 \\ -1 \end{pmatrix} + 2\begin{pmatrix} -3 \\ -2 \\ 4 \end{pmatrix} = 3\mathbf{i} + 2\mathbf{j} + 5\mathbf{k} = \frac{1}{2}(6\mathbf{i} + 4\mathbf{j} + 10\mathbf{k})$

3 $p = 2$, $q = 1$, $r = 2$

4 a $\sqrt{35}$ **b** $2\sqrt{5}$ **c** $\sqrt{3}$ **d** $\sqrt{170}$ **e** $5\sqrt{3}$

5 a $\begin{pmatrix} 7 \\ 1 \\ -1 \end{pmatrix}$ **b** $\begin{pmatrix} -5 \\ 5 \\ -5 \end{pmatrix}$ **c** $\begin{pmatrix} 14 \\ -3 \\ 1 \end{pmatrix}$ **d** $\begin{pmatrix} 8 \\ 4 \\ 4 \end{pmatrix}$ **e** $\begin{pmatrix} 8 \\ -6 \\ 10 \end{pmatrix}$

6 $7\mathbf{i} - 3\mathbf{j} + 2\mathbf{k}$ **7** 6 or –6 **8** $\sqrt{3}$ or $-\sqrt{3}$

9 a i A: $2\mathbf{i} + \mathbf{j} + 4\mathbf{k}$, B: $3\mathbf{i} - 2\mathbf{j} + 4\mathbf{k}$, C: $-\mathbf{i} + 2\mathbf{j} + 2\mathbf{k}$
    ii $-3\mathbf{i} + \mathbf{j} - 2\mathbf{k}$
  b i $\sqrt{14}$ **ii** 3

10 a $-4\mathbf{i} + 3\mathbf{j} - 12\mathbf{k}$ **b** 13 **c** $-\frac{4}{13}\mathbf{i} + \frac{3}{13}\mathbf{j} - \frac{12}{13}\mathbf{k}$

11 a $-6\mathbf{i} + 4\mathbf{j} + 3\mathbf{k}$ **b** $\sqrt{61}$ **c** $-\frac{6}{\sqrt{61}}\mathbf{i} + \frac{4}{\sqrt{61}}\mathbf{j} + \frac{3}{\sqrt{61}}\mathbf{k}$

12 a $\frac{3}{\sqrt{29}}\mathbf{i} - \frac{4}{\sqrt{29}}\mathbf{j} - \frac{2}{\sqrt{29}}\mathbf{k}$ **b** $\frac{\sqrt{2}}{5}\mathbf{i} - \frac{4}{5}\mathbf{j} - \frac{\sqrt{7}}{5}\mathbf{k}$

  c $\frac{\sqrt{5}}{4}\mathbf{i} - \frac{2\sqrt{2}}{4}\mathbf{j} - \frac{\sqrt{3}}{4}\mathbf{k}$

13 a $\overrightarrow{AB} = 4\mathbf{j} - \mathbf{k}$, $\overrightarrow{AC} = 4\mathbf{i} + \mathbf{j} - \mathbf{k}$, $\overrightarrow{BC} = 4\mathbf{i} - 3\mathbf{j}$
  b $|\overrightarrow{AB}| = \sqrt{17}$, $|\overrightarrow{AC}| = 3\sqrt{2}$, $|\overrightarrow{BC}| = 5$
  c scalene

14 a $\overrightarrow{AB} = -2\mathbf{i} - 6\mathbf{j} - 3\mathbf{k}$, $\overrightarrow{AC} = 4\mathbf{i} - 9\mathbf{j} - \mathbf{k}$,
    $\overrightarrow{BC} = 6\mathbf{i} - 3\mathbf{j} + 2\mathbf{k}$
  b $|\overrightarrow{AB}| = 7$, $|\overrightarrow{AC}| = 7\sqrt{2}$, $|\overrightarrow{BC}| = 7$ **c** 45°

15 a i 98.0° **ii** 11.4° **iii** 82.0°
  b i 69.6° **ii** 62.3° **iii** 35.5°
  c i 56.3° **ii** 90° **iii** 146.3°

Online Worked solutions are available in SolutionBank.

16 5.41

17 $\left|\overrightarrow{PQ}\right| = \sqrt{14}, \left|\overrightarrow{QR}\right| = \sqrt{29}, \left|\overrightarrow{PR}\right| = \sqrt{35}$

Let $\theta = \angle PQR$. $14 + 29 - 2\sqrt{406} \cos\theta = 35$
$\Rightarrow \cos\theta = 0.198... \Rightarrow \theta = 78.5°$ (1 d.p.)

Challenge

25.4°

Exercise 7E

1 $\overrightarrow{XY} = \mathbf{b} - \mathbf{a}$ and $\overrightarrow{YZ} = \mathbf{c} - \mathbf{b}$, so $\mathbf{b} - \mathbf{a} = \mathbf{c} - \mathbf{b}$.
Hence $\mathbf{a} + \mathbf{c} = 2\mathbf{b}$.

2 **a** **i** $2\mathbf{r}$ **ii** $\mathbf{r}$
  b Sides of triangle OAB are twice the length of sides of triangle PAQ and angle A is common to both SAS.

3 **a** $\frac{2}{3}\mathbf{a} + \frac{1}{3}\mathbf{b}$
  b $\overrightarrow{AN} = \frac{1}{3}(\mathbf{b} - \mathbf{a}), \overrightarrow{AB} = \mathbf{b} - \mathbf{a}, \overrightarrow{NB} = \frac{2}{3}(\mathbf{b} - \mathbf{a})$
    so $AN:NB = 1:2$.

4 **a** $\frac{3}{5}\mathbf{a} + \frac{2}{5}\mathbf{c}$
  b $\overrightarrow{AP} = -\mathbf{a} + \frac{3}{5}\mathbf{a} + \frac{2}{5}\mathbf{c} = \frac{2}{5}(\mathbf{c} - \mathbf{a})$,
    $\overrightarrow{PC} = \mathbf{c} - (\frac{3}{5}\mathbf{a} + \frac{2}{5}\mathbf{c}) = \frac{3}{5}(\mathbf{c} - \mathbf{a})$ so $AP:PC = 2:3$

5 **a** $\sqrt{26}$ **b** $2\sqrt{2}$ **c** $3\sqrt{2}$
  d $\angle BAC = 56°, \angle ABC = 34°, \angle ACB = 90°$

6 **a** $\overrightarrow{OR} = \mathbf{a} + \frac{1}{3}(\mathbf{b} - \mathbf{a}) = \frac{2}{3}\mathbf{a} + \frac{1}{3}\mathbf{b}$,
    $\overrightarrow{OS} = 3\overrightarrow{OR} = 3(\frac{2}{3}\mathbf{a} + \frac{1}{3}\mathbf{b}) = 2\mathbf{a} + \mathbf{b}$
  b $\overrightarrow{TP} = \overrightarrow{TO} + \overrightarrow{OP} = \mathbf{a} + \mathbf{b}, \overrightarrow{PS} = \overrightarrow{PO} + \overrightarrow{OS} = -\mathbf{a} + 2\mathbf{a} + \mathbf{b}$
    $= \mathbf{a} + \mathbf{b}$
    $\overrightarrow{TP}$ is parallel (and equal) to $\overrightarrow{PS}$ and they have a point, P, in common so T, P and S lie on a straight line.

Challenge:

a $\overrightarrow{PR} = \mathbf{b} - \mathbf{a}, \overrightarrow{PX} = j(\mathbf{b} - \mathbf{a}) = -j\mathbf{a} + j\mathbf{b}$

b $\overrightarrow{ON} = \mathbf{a} + \frac{1}{2}\mathbf{b}, \overrightarrow{PX} = -\mathbf{a} + k(\mathbf{a} + \frac{1}{2}\mathbf{b}) = (k - 1)\mathbf{a} + \frac{1}{2}k\mathbf{b}$

c Coefficients of $\mathbf{a}$ and $\mathbf{b}$ must be the same in both expressions for $\overrightarrow{PX}$
  Coefficients of $\mathbf{a}$: $k - 1 = -j$; Coefficients of $\mathbf{b}$: $j = \frac{1}{2}k$

d Solving simultaneously gives $j = \frac{1}{3}$ and $k = \frac{2}{3}$

e $\overrightarrow{PX} = \frac{1}{3}\overrightarrow{PR}$.
  By symmetry, $\overrightarrow{PX} = \overrightarrow{YR} = \overrightarrow{XY}$, so ON and OM divide PR into 3 equal parts.

Exercise 7F

1 **a** **i** $\left|\overrightarrow{OA}\right| = 9; \left|\overrightarrow{OB}\right| = 9 \Rightarrow \left|\overrightarrow{OA}\right| = \left|\overrightarrow{OB}\right|$

  ii $\overrightarrow{AC} = \begin{pmatrix} 9 \\ 4 \\ 22 \end{pmatrix}, \left|\overrightarrow{AC}\right| = \sqrt{581}; \overrightarrow{BC} = \begin{pmatrix} 6 \\ -4 \\ 23 \end{pmatrix}, \left|\overrightarrow{BC}\right| = \sqrt{581}$

    Therefore $\left|\overrightarrow{AC}\right| = \left|\overrightarrow{BC}\right|$

  b $OACB$ is a kite

2 **a** $\overrightarrow{AB} = 2\mathbf{i} + 3\mathbf{j} - 2\mathbf{k} \Rightarrow \left|\overrightarrow{AB}\right| = \sqrt{17}$
    $\overrightarrow{AC} = 6\mathbf{j} \Rightarrow \left|\overrightarrow{AC}\right| = 6$
    $\overrightarrow{BC} = -2\mathbf{i} + 3\mathbf{j} + 2\mathbf{k} \Rightarrow \left|\overrightarrow{BC}\right| = \sqrt{17}$
    $\left|\overrightarrow{AB}\right| = \left|\overrightarrow{BC}\right|$, so ABC is isosceles.
  b $6\sqrt{2}$
  c (4, 10, 3), (0, 4, 7) or (4, −2, 3)

3 **a** $\overrightarrow{AB} = 4\mathbf{i} - 10\mathbf{j} - 8\mathbf{k} = 2(2\mathbf{i} - 5\mathbf{j} - 4\mathbf{k})$
    $\overrightarrow{CD} = -6\mathbf{i} + 15\mathbf{j} + 12\mathbf{k} = -3(2\mathbf{i} - 5\mathbf{j} - 4\mathbf{k})$
    $\overrightarrow{CD} = -\frac{3}{2}\overrightarrow{AB}$, so AB is parallel to CD
    $AB:CD = 2:3$
  b $ABCD$ is a trapezium

4 $a = \frac{8}{3}, b = -1, c = \frac{3}{2}$

5 $(7, 14, -22), (-7, 14, -22)$ and $\left(\frac{1813}{20}, 14, -22\right)$

6 **a** 18.67 (2 d.p.)   **b** 168.07 (2 d.p.)

7 Let H = point of intersection of OF and AG.
  $\overrightarrow{OH} = r\overrightarrow{OF} = \overrightarrow{OA} + s\overrightarrow{AG}$
  $\overrightarrow{OF} = \mathbf{a} + \mathbf{b} + \mathbf{c}, \overrightarrow{AG} = -\mathbf{a} + \mathbf{b} + \mathbf{c}$
  So $r(\mathbf{a} + \mathbf{b} + \mathbf{c}) = \mathbf{a} + s(-\mathbf{a} + \mathbf{b} + \mathbf{c})$
  $r = 1 - s = s \Rightarrow r = s = \frac{1}{2}$, so $\overrightarrow{OH} = \frac{1}{2}\overrightarrow{OF}$ and $\overrightarrow{AH} = \frac{1}{2}\overrightarrow{AG}$.

8 Show that $\overrightarrow{FP} = \frac{2}{3}\mathbf{a}$ (multiple methods possible)
  Show that $\overrightarrow{PE} = \frac{1}{3}\mathbf{a}$ (multiple methods possible)
  Therefore FP and PE are parallel, so P lies on FE
  $FP:PE = 2:1$

Challenge

1 $p = \frac{24}{11}, q = \frac{32}{11}, r = -4$

2 $\overrightarrow{OM} = \frac{1}{2}\mathbf{a} + \mathbf{b} + \mathbf{c}, \overrightarrow{BN} = \mathbf{a} - \mathbf{b} + \frac{1}{2}\mathbf{c}, \overrightarrow{AF} = -\mathbf{a} + \mathbf{b} + \mathbf{c}$
  Let $\overrightarrow{OM}$ and $\overrightarrow{AF}$ intersect at X: $\overrightarrow{AX} = r\overrightarrow{AF} = r(-\mathbf{a} + \mathbf{b} + \mathbf{c})$
  $\overrightarrow{OX} = s\overrightarrow{OM} = s\left(\frac{1}{2}\mathbf{a} + \mathbf{b} + \mathbf{c}\right)$ for scalars r and s
  $\overrightarrow{OX} = \overrightarrow{OA} + \overrightarrow{AX} = \mathbf{a} + r(-\mathbf{a} + \mathbf{b} + \mathbf{c})$
  $\Rightarrow s\left(\frac{1}{2}\mathbf{a} + \mathbf{b} + \mathbf{c}\right) = \mathbf{a} + r(-\mathbf{a} + \mathbf{b} + \mathbf{c})$
  Comparing coefficients in $\mathbf{a}$, $\mathbf{b}$ and $\mathbf{c}$ gives $r = s = \frac{2}{3}$
  Let $\overrightarrow{BN}$ and $\overrightarrow{AF}$ intersect at Y: $\overrightarrow{AY} = p\overrightarrow{AF} = p(-\mathbf{a} + \mathbf{b} + \mathbf{c})$
  $\overrightarrow{BY} = q\overrightarrow{BN} = q\left(\mathbf{a} - \mathbf{b} + \frac{1}{2}\mathbf{c}\right)$ for scalars p and q
  $\overrightarrow{BY} = \overrightarrow{BA} + \overrightarrow{AY} = \mathbf{a} - \mathbf{b} + p(-\mathbf{a} + \mathbf{b} + \mathbf{c})$
  $\Rightarrow q\left(\mathbf{a} - \mathbf{b} + \frac{1}{2}\mathbf{c}\right) = \mathbf{a} - \mathbf{b} + p(-\mathbf{a} + \mathbf{b} + \mathbf{c})$
  Comparing coefficients in $\mathbf{a}$, $\mathbf{b}$ and $\mathbf{c}$ gives $p = \frac{1}{3}, q = \frac{2}{3}$
  $\overrightarrow{AX} = \frac{2}{3}\overrightarrow{AF}, \overrightarrow{AY} = \frac{1}{3}\overrightarrow{AF}$
  So the line segments OM and BN trisect the diagonal AF.

Exercise 7G

1 **a** **i** $\overrightarrow{OA} = 3\mathbf{i} - \mathbf{j}, \overrightarrow{OB} = 4\mathbf{i} + 5\mathbf{j}, \overrightarrow{OC} = -2\mathbf{i} + 6\mathbf{j}$
   ii $\mathbf{i} + 6\mathbf{j}$ **iii** $-5\mathbf{i} + 7\mathbf{j}$
   b **i** $\sqrt{40} = 2\sqrt{10}$ **ii** $\sqrt{37}$ **iii** $\sqrt{74}$

2 **a** $-\mathbf{i} + 5\mathbf{j}$ or $\begin{pmatrix} -1 \\ 5 \end{pmatrix}$

   b **i** 5   **ii** $\sqrt{13}$ **iii** $\sqrt{26}$

3 **a** $-\mathbf{i} - 9\mathbf{j}$ or $\begin{pmatrix} -1 \\ -9 \end{pmatrix}$

   b **i** $\sqrt{82}$ **ii** 5   **iii** $\sqrt{61}$

4 **a** $-2\mathbf{a} + 2\mathbf{b}$ **b** $-3\mathbf{a} + 2\mathbf{b}$ **c** $-2\mathbf{a} + \mathbf{b}$

5 $\begin{pmatrix} 7 \\ 9 \end{pmatrix}$ or $\begin{pmatrix} 9 \\ 3 \end{pmatrix}$

6 **a** $2\mathbf{i} + 8\mathbf{j}$ **b** $2\sqrt{17}$

7 $\dfrac{3\sqrt{5}}{5}$

Challenge

$\overrightarrow{OB} = 2\mathbf{i} + 3\mathbf{j}$ or $\overrightarrow{OB} = \dfrac{46}{13}\mathbf{i} + \dfrac{9}{13}\mathbf{j}$

Exercise 7H

1 $2\sqrt{21}$ **2** $7\sqrt{3}$
3 **a** $\sqrt{14}$ **b** 15   **c** $5\sqrt{2}$ **d** $\sqrt{30}$
4 $k = 5$ or $k = 9$ **5** $k = 10$ or $k = -4$

Challenge

a $(1, -3, 4), (1, -3, -2), (7, 3, 4), (7, 3, -2), (7, -3, -2)$
b $6\sqrt{5}$

Exercise 7I

1 **a** $\mathbf{r} = \begin{pmatrix} 6 \\ 5 \\ -1 \end{pmatrix} + \lambda \begin{pmatrix} 2 \\ -3 \\ -1 \end{pmatrix}$ **b** $\mathbf{r} = \begin{pmatrix} 2 \\ 5 \\ 0 \end{pmatrix} + \lambda \begin{pmatrix} 1 \\ 1 \\ 1 \end{pmatrix}$

   c $\mathbf{r} = \begin{pmatrix} -7 \\ 6 \\ 2 \end{pmatrix} + \lambda \begin{pmatrix} 3 \\ 1 \\ 2 \end{pmatrix}$ **d** $\mathbf{r} = \begin{pmatrix} 2 \\ 0 \\ 4 \end{pmatrix} + \lambda \begin{pmatrix} -3 \\ 2 \\ 1 \end{pmatrix}$

   e $\mathbf{r} = \begin{pmatrix} 6 \\ -11 \\ 2 \end{pmatrix} + \lambda \begin{pmatrix} 0 \\ 5 \\ -2 \end{pmatrix}$

2 **a** **i** $2\mathbf{i} + 7\mathbf{j} - 3\mathbf{k}$ **ii** $\mathbf{r} = (3\mathbf{i} - 4\mathbf{j} + 2\mathbf{k}) + \lambda(2\mathbf{i} + 7\mathbf{j} - 3\mathbf{k})$
   b **i** $2\mathbf{i} - 3\mathbf{j} + 4\mathbf{k}$ **ii** $\mathbf{r} = (2\mathbf{i} + \mathbf{j} - 3\mathbf{k}) + \lambda(2\mathbf{i} - 3\mathbf{j} + 4\mathbf{k})$
   c **i** $-3\mathbf{i} - \mathbf{j} - 2\mathbf{k}$ **ii** $\mathbf{r} = (\mathbf{i} - 2\mathbf{j} + 4\mathbf{k}) + \lambda(-3\mathbf{i} - \mathbf{j} - 2\mathbf{k})$
   d **i** $\begin{pmatrix} -5 \\ 4 \\ -3 \end{pmatrix}$ **ii** $\mathbf{r} = \begin{pmatrix} 3 \\ -1 \\ 4 \end{pmatrix} + \lambda \begin{pmatrix} -5 \\ 4 \\ -3 \end{pmatrix}$
   e **i** $\begin{pmatrix} -6 \\ 4 \\ 1 \end{pmatrix}$ **ii** $\mathbf{r} = \begin{pmatrix} 4 \\ -2 \\ 3 \end{pmatrix} + \lambda \begin{pmatrix} -6 \\ 4 \\ 1 \end{pmatrix}$

3 **a** $\mathbf{r} = \begin{pmatrix} 4 \\ -3 \\ 8 \end{pmatrix} + \lambda \begin{pmatrix} 0 \\ 0 \\ 1 \end{pmatrix}$

4 **a** $p = 1, q = 10$ **b** $p = -6\frac{1}{2}, q = -21$
   c $p = -19, q = -15$

5 Direction of l_1: $\begin{pmatrix} 2 \\ -3 \\ 4 \end{pmatrix}, \overrightarrow{AB} = \begin{pmatrix} -2 \\ 3 \\ -4 \end{pmatrix} = -\begin{pmatrix} 2 \\ -3 \\ 4 \end{pmatrix}$, so parallel

6 $\overrightarrow{AB} = \begin{pmatrix} 6 \\ 3 \\ -3 \end{pmatrix}, \overrightarrow{BC} = \begin{pmatrix} 6 \\ 3 \\ -3 \end{pmatrix}$ same direction and a point in common

7 $\begin{pmatrix} 3 \\ -1 \\ 8 \end{pmatrix} - \begin{pmatrix} 1 \\ 7 \\ -2 \end{pmatrix} = \begin{pmatrix} 2 \\ -8 \\ 10 \end{pmatrix} = 2\begin{pmatrix} 1 \\ -4 \\ 5 \end{pmatrix}; \begin{pmatrix} 10 \\ 4 \\ 0 \end{pmatrix} - \begin{pmatrix} 1 \\ 7 \\ -2 \end{pmatrix} = \begin{pmatrix} 9 \\ -3 \\ 2 \end{pmatrix}$
   so not collinear
8 $a = 2.5, b = -2$
9 $\mathbf{r} = (2\mathbf{i} - 7\mathbf{j} + 16\mathbf{k}) + \lambda(2\mathbf{i} - 4\mathbf{j} + \mathbf{k})$
10 **a** $a = 14, b = -2$ **b** $X(9, 9, -10)$
11 $AB = 9$

12 $B\begin{pmatrix} 11 \\ 3 \\ -2 \end{pmatrix}$

13 **a** $A\begin{pmatrix} -2 \\ 4 \\ 7 \end{pmatrix}, B\begin{pmatrix} 1 \\ 1 \\ 10 \end{pmatrix}$ **b** $\mathbf{r} = \begin{pmatrix} 0 \\ 2 \\ 3 \end{pmatrix} + \lambda \begin{pmatrix} 1 \\ -1 \\ 1 \end{pmatrix}$

   c $C\begin{pmatrix} 1 \\ 1 \\ 4 \end{pmatrix}, D\begin{pmatrix} -1 \\ 3 \\ 2 \end{pmatrix}$ so P is midpoint

14 **a** $A(10, 9, 8)$
   b Tightrope will bow in the middle with acrobat's weight

Exercise 7J

1 **a** The two lines do meet at the point $(3, 1, 10)$
   b The lines do not meet.
   c The two lines do meet at the point $(0, 1\frac{1}{2}, 4\frac{1}{2})$

2 l_1 and l_2 meet when $\lambda = 4$ and $\mu = -2$
   coordinates of point of intersection $(-2, -4, 15)$
3 No solution for λ and μ
4 **a** $(p = 3)$ **b** $(2, 5, -3)$
5 $(6, 1, -1)$
6 **a** Solve $\begin{pmatrix} -1 \\ 3 \\ 2 \end{pmatrix} \cdot \begin{pmatrix} q \\ 2 \\ -1 \end{pmatrix} = 0$ **b** $p = -2$

   c $(4, 14, -4)$ **d** $\begin{pmatrix} -1 \\ 29 \\ 6 \end{pmatrix}$

Exercise 7K

1 $\dfrac{9}{2}$

2 **a** 2   **b** 17   **c** -6 **d** 20   **e** 0
3 **a** $55.5°$ **b** $94.8°$ **c** $87.4°$ **d** $79.0°$
   e $100.9°$ **f** $53.7°$ **g** $74.3°$ **h** $70.5°$
4 **a** -10 **b** 5   **c** $2\frac{3}{5}$
   d $-2\frac{1}{2}$ **e** -5 or 2
5 **a** $32.9°$ **b** $117.8°$
6 **a** $20.5°$ **b** $109.9°$
7 $\dfrac{2\sqrt{2}}{3}$

8 Use $\begin{pmatrix} 1 \\ 3 \\ 0 \end{pmatrix} \cdot \begin{pmatrix} 0 \\ 1 \\ \lambda \end{pmatrix} = 3 = \sqrt{10}\sqrt{\lambda^2 + 1} \cos 60°$

9 **a** $\mathbf{i} + 2\mathbf{j} + \mathbf{k}$ **b** $3\mathbf{i} + 2\mathbf{j} + 3\mathbf{k}$ **c** $3\mathbf{i} + 2\mathbf{j} + 4\mathbf{k}$
10 $64.7°, 64.7°, 50.6°$
11 **a** $|\overrightarrow{AB}| = \sqrt{33}, |\overrightarrow{BC}| = \sqrt{173}$
   b $29.1°$
12 **a** $\cos\theta = \dfrac{26}{27}$ **b** area $= \dfrac{1}{2} \times 9 \times 3 \times \dfrac{\sqrt{53}}{27}$
13 Let $\overrightarrow{OA} = \mathbf{a}, \overrightarrow{OP} = \mathbf{p}$; then $\overrightarrow{OB} = \mathbf{b} = -\mathbf{a}$ and find scalar product

14 a $\overrightarrow{CA} = \begin{pmatrix} -1 \\ 0 \\ -4 \end{pmatrix}$, $\overrightarrow{CB} = \begin{pmatrix} -4 \\ 5 \\ 6 \end{pmatrix}$

b $\dfrac{3\sqrt{101}}{2}$

c $(9, -6, -6)$, $(1, 4, 6)$, $(3, 4, 6)$

d $3\sqrt{101}$

15 a $\overrightarrow{PQ} = \begin{pmatrix} -3 \\ 6 \\ -2 \end{pmatrix}$, $\overrightarrow{QR} = \begin{pmatrix} 2 \\ -2 \\ 9 \end{pmatrix}$; scalar product = 0

b centre $(0.5, 1, 0.5)$, radius $= \dfrac{\sqrt{138}}{2}$

Challenge

1 $\mathbf{a.b} = |\mathbf{a}||\mathbf{b}| \cos\theta$, $\mathbf{b.a} = |\mathbf{b}||\mathbf{a}|\cos\theta$ so $\mathbf{a.b} = \mathbf{b.a}$

2 a i $\mathbf{a.(b + c)} = |\mathbf{a}||\mathbf{b + c}| \cos\theta$, but $\cos\theta = \dfrac{PQ}{|\mathbf{b + c}|}$

so $\mathbf{a.(b + c)} = |\mathbf{a}| \times PQ$

ii $\mathbf{a.b} = |\mathbf{a}||\mathbf{b}| \cos\alpha$, but $\cos\alpha = \dfrac{PR}{|\mathbf{b}|}$

so $\mathbf{a.b} = |\mathbf{a}| \times PR$

iii $\mathbf{a.c} = |\mathbf{a}||\mathbf{c}| \cos\beta$, but $\cos\beta = \dfrac{MN}{|\mathbf{c}|} = \dfrac{RQ}{|\mathbf{c}|}$

so $\mathbf{a.c} = |\mathbf{a}| \times RQ$

b $\mathbf{a.(b + c)} = |\mathbf{a}| \times PQ = |\mathbf{a}| \times (PR + RQ) = (|\mathbf{a}| \times PR)$ $+ (|\mathbf{a}| \times RQ) = \mathbf{a.b} + \mathbf{a.c}$; so $\mathbf{a.(b + c)} = \mathbf{a.b} + \mathbf{a.c}$

Chapter review 7

1 a $2\sqrt{10}$ newtons **b** 18° to the left

2 a 108° **b** 9.49 km h⁻¹

3 a $\mathbf{b} - \dfrac{3}{5}\mathbf{a}$ **b** $\mathbf{b} - 4\mathbf{a}$ **c** $\dfrac{8}{5}\mathbf{a} - \mathbf{b}$ **d** $3\mathbf{a} - \mathbf{b}$

4 1.25

5 a $\begin{pmatrix} 12 \\ -1 \end{pmatrix}$ **b** $\begin{pmatrix} -18 \\ 5 \end{pmatrix}$ **c** $\begin{pmatrix} 49 \\ 13 \end{pmatrix}$

6 a $p = -1.5$ **b** $\mathbf{i} - 1.5\mathbf{j}$

7 $p = 8.6$, $q = 12.3$

8 ± 6

9 a $\dfrac{3}{5}\mathbf{a} + \dfrac{2}{5}\mathbf{b}$ **b** $\dfrac{2}{5}\mathbf{b}$

c $\overrightarrow{AB} = \mathbf{b} - \mathbf{a}$, $\overrightarrow{AN} = \dfrac{2}{5}(\mathbf{b} - \mathbf{a})$ so $AN:NB = 2:3$

10 $\sqrt{22}$ **11** $a = 5$ or $a = 6$

12 $|\overrightarrow{AB}| = 5\sqrt{2} \Rightarrow 9 + t^2 + 25 = 50 \Rightarrow t^2 = 16 \Rightarrow t = 4$

$6\mathbf{i} - 8\mathbf{j} - \dfrac{5}{2}t\mathbf{k} = 6\mathbf{i} - 8\mathbf{j} - 10\mathbf{k} = -2(-3\mathbf{i} + 4\mathbf{j} + 5\mathbf{k}) = -2\overrightarrow{AB}$

So $\overrightarrow{AB}$ is parallel to $6\mathbf{i} - 8\mathbf{j} - \dfrac{5}{2}t\mathbf{k}$

13 a $\overrightarrow{PQ} = -3\mathbf{i} - 8\mathbf{j} + 3\mathbf{k}$, $\overrightarrow{PR} = -3\mathbf{i} - 9\mathbf{j} + 8\mathbf{k}$, $\overrightarrow{QR} = -\mathbf{j} + 5\mathbf{k}$

b 20.0

14 a $\overrightarrow{DE} = 4\mathbf{i} + 3\mathbf{j} + 4\mathbf{k}$, $\overrightarrow{EF} = -3\mathbf{i} - 4\mathbf{j} + 4\mathbf{k}$, $\overrightarrow{FD} = -\mathbf{i} + \mathbf{j} + 8\mathbf{k}$

b $|\overrightarrow{DE}| = \sqrt{41}$, $|\overrightarrow{EF}| = \sqrt{41}$, $|\overrightarrow{FD}| = \sqrt{66}$ **c** isosceles

15 a $\overrightarrow{PQ} = 9\mathbf{i} - 4\mathbf{j}$, $\overrightarrow{PR} = 7\mathbf{i} + \mathbf{j} - 3\mathbf{k}$, $\overrightarrow{QR} = -2\mathbf{i} + 5\mathbf{j} - 3\mathbf{k}$

b $|\overrightarrow{PQ}| = \sqrt{97}$, $|\overrightarrow{PR}| = \sqrt{59}$, $|\overrightarrow{QR}| = \sqrt{38}$ **c** 51.3°

16 31.5°

17 184 (3 s.f.)

18 a $(2, -7, -2)$ **b** rhombus **c** 36.1

19 $\overrightarrow{PQ} = \dfrac{1}{2}(\mathbf{a} + \mathbf{b} - \mathbf{c})$, $\overrightarrow{RS} = \dfrac{1}{2}(-\mathbf{a} + \mathbf{b} + \mathbf{c})$, $\overrightarrow{TU} = \dfrac{1}{2}(\mathbf{a} - \mathbf{b} + \mathbf{c})$

Let $\overrightarrow{PQ}$, $\overrightarrow{RS}$ and $\overrightarrow{TU}$ intersect at X: $\overrightarrow{PX} = r\overrightarrow{PQ} = \dfrac{r}{2}(\mathbf{a} + \mathbf{b} - \mathbf{c})$

$\overrightarrow{RX} = s\overrightarrow{RS} = \dfrac{s}{2}(-\mathbf{a} + \mathbf{b} + \mathbf{c})$

$\overrightarrow{TX} = t\overrightarrow{TU} = \dfrac{t}{2}(\mathbf{a} - \mathbf{b} + \mathbf{c})$ for scalars r, s and t

$\overrightarrow{RX} = \overrightarrow{RO} + \overrightarrow{OP} + \overrightarrow{PX} = \dfrac{1}{2}(-\mathbf{a} + \mathbf{c}) + \dfrac{r}{2}(\mathbf{a} + \mathbf{b} - \mathbf{c})$

$\Rightarrow \dfrac{s}{2}(-\mathbf{a} + \mathbf{b} + \mathbf{c}) = \dfrac{1}{2}(-\mathbf{a} + \mathbf{c}) + \dfrac{r}{2}(\mathbf{a} + \mathbf{b} - \mathbf{c})$

Comparing coefficients in **a**, **b** and **c** gives $r = s = \dfrac{1}{2}$

$\overrightarrow{TX} = \overrightarrow{TO} + \overrightarrow{OP} + \overrightarrow{PX} = \dfrac{1}{2}(-\mathbf{b} + \mathbf{c}) + \dfrac{1}{4}(\mathbf{a} + \mathbf{b} - \mathbf{c})$

$\Rightarrow \dfrac{t}{2}(\mathbf{a} - \mathbf{b} + \mathbf{c}) = \dfrac{1}{4}(\mathbf{a} - \mathbf{b} + \mathbf{c})$

Comparing coefficients in **a**, **b** and **c** gives $t = \dfrac{1}{2}$

So the line segments PQ, RS and TU meet at a point and bisect each other.

20 $b = 1$ or $\dfrac{17}{3}$

21 a Air resistance acts in opposition to the motion of the BASE jumper. The motion downwards will be greater than the motion in the other directions.

b $(16\mathbf{i} + 13\mathbf{j} - 40\mathbf{k})$ N **c** 20 seconds

22 a $\mathbf{r} = (\mathbf{i} - \mathbf{j} + 3\mathbf{k}) + \lambda(3\mathbf{j} - \mathbf{k})$ **b** $\mathbf{i} + \mathbf{j} + \dfrac{7}{3}\mathbf{k}$

23 $\mathbf{r} = (2\mathbf{i} + 3\mathbf{j} - 4\mathbf{k}) + \lambda(2\mathbf{j} + 3\mathbf{k})$

24 $7\mathbf{i} + 4\mathbf{j} - 5\mathbf{k}$ lies on l when $\lambda = 2$

$9\mathbf{i} + 3\mathbf{j} - 6\mathbf{k} = 3(3\mathbf{i} + \mathbf{j} - 2\mathbf{k})$ so parallel

25 a $3\mathbf{i} + 4\mathbf{j} + 5\mathbf{k}$, $\mathbf{i} + \mathbf{j} + 4\mathbf{k}$

b $\dfrac{\overrightarrow{ML} \cdot \overrightarrow{MN}}{|\overrightarrow{ML}||\overrightarrow{MN}|} = \dfrac{27}{5\sqrt{2} \, 3\sqrt{2}} = \dfrac{9}{10}$

26 a $\mathbf{r} = \begin{pmatrix} 9 \\ -2 \\ 1 \end{pmatrix} + \mu\begin{pmatrix} -3 \\ 4 \\ 5 \end{pmatrix}$ **b** $p = 6$, $q = 11$

c 39.8° **d** $\dfrac{36}{5}\mathbf{i} + \dfrac{2}{5}\mathbf{j} + 4\mathbf{k}$

27 a $\mathbf{r} = \begin{pmatrix} 1 \\ 2 \\ -3 \end{pmatrix} + \mu\begin{pmatrix} 4 \\ -2 \\ 0 \end{pmatrix}$ **b** $\mu = -3$

c 53.4° **d** $\dfrac{\sqrt{145}}{5}$

28 a $\mathbf{r_1.r_2} = 0$, therefore vectors are perpendicular

b $5\mathbf{i} - \mathbf{k}$ **c** $l_1 : \lambda = -3$ **d** 1.5 km

29 a intersect when $\lambda = 3$, $\mu = -2$

b $\begin{pmatrix} 7 \\ 4 \\ -6 \end{pmatrix}$ **c** $\dfrac{4\sqrt{10}}{15}$

30 a $a = 11$, $b = 7$ **b** $P(5, 9, 4)$ **c** $\sqrt{122}$

31 a $\overrightarrow{AB} = \begin{pmatrix} -1 \\ -1 \\ 2 \end{pmatrix}$ **b** $\mathbf{r} = \begin{pmatrix} 6 \\ 3 \\ 4 \end{pmatrix} + \lambda\begin{pmatrix} -1 \\ -1 \\ 2 \end{pmatrix}$

c $\begin{pmatrix} 7.5 \\ 4.5 \\ 1 \end{pmatrix}$

32 a meet when $\lambda = 2$, $\mu = 6$, $(7, 0, 2)$

b 80.4°

c lies on l_2 when $\lambda = 1$

d 2.42

33 a intersect at $(180, -5, 7)$

b pass through same point but not necessarily at the same time

Review exercise 2

1 a $-2\sin^3 t \cos t$ **b** $y = -\dfrac{1}{2}x + 2$

c $y = \dfrac{8}{4 + x^2}$

$x \geqslant 0$ is the domain of the function.

2 **a** $y = -9x + 8$ **b** $y = \dfrac{x}{2x-1}$

3 $7x + 2y - 2 = 0$

4 **a** $\dfrac{dy}{dx} = \dfrac{\cos x}{\sin y}$

   b Stationary points at $\left(\dfrac{\pi}{2}, \dfrac{2\pi}{3}\right)$ and $\left(\dfrac{\pi}{2}, \dfrac{-2\pi}{3}\right)$ only in the given range.

5 **a** $\dfrac{dV}{dr} = 4\pi r^2$ **b** $\dfrac{dr}{dt} = \dfrac{250}{\pi(2t+1)^2 r^2}$

6 **a** $\dfrac{dy}{dx} = -\dfrac{x+2y}{x}$ **b** $y = -3x + 4$

7 **a** $k = \dfrac{1}{3}$ **b** $a = 4$

   c $y = \dfrac{2}{5}x + 1$

8 **a** $x = \sec^2 t = (\sec t)^2$ $\dfrac{dx}{dt} = 2(\sec t)(\sec t \tan t)$
   $= 2\sec^2 t \tan t$

   b 2

9 $a = 1$

10 **a** 1   **b** $\dfrac{2\pi}{15}$

11 $\dfrac{\pi}{8}(\pi + 2)$

12 16   **13** $\dfrac{2}{3} - \dfrac{3\sqrt{3}}{8}$

14 $\dfrac{1}{9}(2e^3 + 10)$

15 **a** $\dfrac{5x+3}{(2x-3)(x-2)} \equiv \dfrac{3}{2x-3} + \dfrac{1}{x+2}$ **b** $\ln 54$

16 $\dfrac{e^{-x}}{5}(2\sin 2x - \cos 2x) + c$

17 **a** $\dfrac{2x-1}{(x-1)(2x-3)} \equiv \dfrac{-1}{x-1} + \dfrac{4}{2x-3}$

   b $y = \dfrac{A(2x-3)^2}{(x-1)}$ **c** $y = \dfrac{10(2x-3)^2}{(x-1)}$

18 **a** $\dfrac{3k}{16\pi^2 r^5}$ **b** $r = \left[\dfrac{9k}{8\pi^2}t + A'\right]^{\frac{1}{6}}$

19 **a** Rate in = 20, rate out = $-kV$. So $\dfrac{dV}{dt} = 20 - kV$

   b $A = \dfrac{20}{k}$ and $B = -\dfrac{20}{k}$

   c $108\,\text{cm}^3$ (3 s.f.)

20 **a** $\dfrac{dC}{dt} = -kC$, because k is the constant of proportionality. The negative sign and $k > 0$ indicates rate of decrease.

   b $C = Ae^{-kt}$ **c** $k = \dfrac{1}{4}\ln 10$

21 -4.5

22 $\sqrt{7}$

23 12

24 $130.3°$

25 **a** $10\mathbf{i} - 5\mathbf{j} - 2\mathbf{k}$ **b** $\dfrac{10}{\sqrt{129}}\mathbf{i} - \dfrac{5}{\sqrt{129}}\mathbf{j} - \dfrac{2}{\sqrt{129}}\mathbf{k}$

   c $100.1°$ **d** Not parallel: $\overrightarrow{PQ} \neq m\overrightarrow{AB}$.

26 $k = 2$ **27** $p = -2, q = -8, r = -4$

28 $5\sqrt{14}$

29 $a = 3, b = 13, \mathbf{r} = \begin{pmatrix} 1 \\ -1 \\ 3 \end{pmatrix} + \lambda \begin{pmatrix} 2 \\ 4 \\ 5 \end{pmatrix}$ or any equivalent

30 **a** As the solution $\lambda = -2, \mu = -3$ satisfies all three equations, the lines *do* meet.

   b $(3, 1, -2)$ is point of intersection.

   c $\dfrac{5}{9}\sqrt{3}$

31 **a** $a = 18$ $b = 9$

   b $(6, 10, 16)$

   c $14\sqrt{2}$

32 **a** Lines do not intersect.

   b Unlikely that the shark will not adjust course to intercept flounder.

Challenge

1 **a** $\dfrac{dy}{dx} = \dfrac{3x^2 - y^2}{2xy + 1}$

   b $\dfrac{d^2y}{dx^2} = \dfrac{6x - 4y\dfrac{dy}{dx} - 2x\left(\dfrac{dy}{dx}\right)^2}{2xy + 1}$

   c $\dfrac{11}{5}$ and $-\dfrac{404}{125}$

2 $\ln\dfrac{3}{2}$

3 $\left(-\dfrac{1}{2}, -1, \dfrac{1}{2}\right)$, radius $= \sqrt{\dfrac{13}{2}}$

Exam Practice

1 Assumption: if $n^2 + 1$ is even then can be even.

   Let $n = 2k$ where k is an integer.

   Then $(2k)^2 + 1 = 4k^2 + 1$.

   But $4k^2$ is even, and so $4k^2 + 1$ must be odd.

   This contradicts our assumption, and so if $n^2 + 1$ is even then n must be odd.

2 **a** $\dfrac{dy}{dx} = \dfrac{-x - 2y}{2x + 2y}$

   b $\left(-\dfrac{2}{\sqrt{3}}, \dfrac{1}{\sqrt{3}}\right), \left(\dfrac{2}{\sqrt{2}}, -\dfrac{1}{\sqrt{3}}\right)$

3 **a** $\left(-\dfrac{1}{2}, -\dfrac{1}{2e}\right)$

   b $\dfrac{e^2}{4}(2e^3 - 1)$

4 **a** $y = 2x^2 - 1$

   b $-1 \leqslant x \leqslant 1, -1 \leqslant y \leqslant 1$

   c $4\cos t$

   d $y = 4x - 3$

5 **a** $1 + 6x + 27x^2 + 108x^3$

   b 1.062808

6 **a** $H = 5 + 35e^{-20t}$

   b $-700e^{-20t}, -0.0318$

   c 5

7 **a** $\mathbf{r} = 2\mathbf{i} + \mathbf{j} + 3\mathbf{k} + (3\mathbf{i} - 3\mathbf{j} - 2\mathbf{k})t$

   c $129.8°$

   d $(11, -8, -3), (-7, 10, 9)$

8 $\dfrac{\pi}{6}$

Online Worked solutions are available in SolutionBank.

INDEX